The Social Direction of Evolution

An Outline of the Science of Eugenics

William E. Kellicott

I

THE SOURCES AND AIMS OF THE SCIENCE OF EUGENICS

"Bravas to all impulses sending sane children to the next age!"

Eugenics has been defined as "the science of being well born." In the words of Sir Francis Galton, who may fairly be claimed as the founder of this newest of sciences, "Eugenics is the study of the agencies under social control, that may improve or impair the racial qualities of future generations, either physically or mentally."

The idea of definitely undertaking to improve the innate characteristics of the human race has been expressed repeatedly through centuries—fancifully, seriously, hopefully, and now scientifically. Since the times of Theognis and of Plato the student of animate Nature has been aware of the possibility of the degradation or of the elevation of the human race-characters. The conditions under which life exists gradually change: the customs and ideals of societies change rapidly. Times inevitably come when, if we are to maintain or to advance our racial position, we find it necessary to change in an adaptive way our attitude toward these changing social relations and conditions of life. If we neglect to do this we go down in the racial struggle, as history so clearly and so repeatedly warns us.

In the opinion of many biologists and sociologists such a time has now arrived. The suspension of many forms of natural selection in human society, the currency of the "rabbit theory" of racial prosperity—based upon the idea of mere numerical increase of the population, the complacent disregard of the increase of the pauper, insane, and criminal elements of our population, the dearth of individuals of high ability—even of competent workmen, all are resulting in evil and will result disastrously unless deliberately controlled. It is hoped that this control, though at first conscious, "artificial," may later become fixed as an element of social custom and conscience and thus operate automatically and the more effectively. The result will be not only the restoration of our race to its original vigor, mental and physical, but further the carrying on of the race to a surpassing vigor and supremacy.

The aim of Eugenics is the production of a more healthy, more vigorous, more able humanity. Again in the words of Galton "The aim of Eugenics is to represent each class ... by its best specimens; that done to leave them to work out their common civilization in their own way.... To bring as many influences as can be reasonably employed to cause the useful classes in the community to contribute more than their present proportion to the next generation"; and further, we might add, to cause the useless, vicious classes to contribute to the next generation less than their present proportion.

With this definition of Eugenics and preliminary statement of its aims before us we may proceed to a somewhat fuller statement of the facts within this field. First let us consider the relation of the science of Eugenics to its parent sciences, biology and sociology, then after mentioning some of the steps in the development of the present eugenic movement, we may describe some of the conditions which give us human beings pause and lead us to appreciate the necessity for a reconsideration of much that enters into our present social organization and conduct.

Shortly before the publication of "The Origin of Species," Darwin was asked by Alfred Russell Wallace whether he proposed to include any reference to the evolution of man. Darwin's reply was: "You ask whether I shall discuss man. I think I shall avoid the whole subject, as so surrounded with prejudices; though I fully admit that it is the highest and most interesting problem for the naturalist." This prejudice which Darwin knew would preclude a just consideration of the subject of man's origin and evolution, grew out of the former and long current conception of the position occupied by man in the whole scheme of Nature—of "Man's Place in Nature."

This conception, happily obsolete now among thinkers, though occasionally seen lurking in out of the way corners shaded from the light of modern philosophy and science, placed Man and the rest of the universe in separate categories. Man was one, all the rest another. It was for Man's benefit or pleasure that the rains descended, that the corn grew and ripened, that the sun shone, the birds sang, the landscape was spread before the view. For Man's warning or punishment the lightning struck, comets appeared, disease ravaged, insects tormented and destroyed. It was certainly very natural that Man should regard himself as a thing apart, particularly since he was able to control and to regulate Nature, and to take tribute from her so extensively. But the scientist regarded man differently; from him the world learned to recognize man as an integral factor in Nature—as one with Nature, possessing the same structures, performing the same activities, as other animals; subject to much the same control and with much the same purposes in life and in Nature as other living things. There is to-day no necessity to enlarge upon this view. As Ray Lankester puts it: "Man is held to be a part of Nature; a being, resulting from and driven by the one great nexus of mechanism which we call Nature."

But the echoes of the older naïve view of Man and his Nature sounded long after the rational scientific conception had become dominant. It is not so very long ago that psychology was little more than human psychology; nor has sociology long since gone outside the purely human for explanations of the facts of human society. Nowadays, however, psychology has a firm comparative basis and sociology finds much that is illuminating and helpful in the purely biological aspects of the human animal. Very naturally, then, we have had social science studying man as Man, with a capital M: biological science studying man as a natural animal.

But now that modern trend of scientific synthesis which has brought forth a Physical-Chemistry and a Chemical-Physiology and a Bio-Chemistry, is combining the purely social and the purely biological studies of man into a new Bio-Sociology. And as one phase of this new partnership we have the subject of Eugenics—the science of racial integrity and progress, built upon the overlapping fields of Biology and Sociology.

We can trace the idea, perhaps better the hope, of Eugenics from the modern times of ancient Greece. Plato laid stress upon the idea of the "purification of the State." In his Republic he pointed out that the quality of the herd or flock could be maintained only by breeding from the best, consciously selected for that purpose by the shepherd, and by the destruction of the weaklings; and that when one was concerned with the quality of his hunting dogs or horses or pet birds, he was careful to utilize this knowledge. He drew attention to the necessity in the State for a functionary corresponding to the shepherd to weed out the undesirables and to prevent them from multiplying their kind. Plato stated clearly the essential idea of the inheritance of individual qualities and the danger to the State of a large and increasing body of degenerates and defectives. He called upon the legislators to purify the State. But the legislators paid no heed. The able-bodied and able-minded continued to be sacrificed to the God of War; the degenerates and defectives—not fit to fight—were the ones left at home to become parents of the next generation. And to-day Greece remains an awful warning.

We cannot describe or even enumerate the wrecks of the many plans for race improvement that are strewn from Plato to our day. Sporadic, emotional, visionary, often it must be confessed suggested by possibilities of material gain to the "leader"—they have all passed. They failed because they were unscientific; because there was available no solid foundation of determined fact upon which to build. One need suggest only the Oneida Community, as it was originally planned, or the Parisian society of *L'Elite* —in both of which the selection of mates was to be carefully controlled—or some of the fantasies of Bernard Shaw, to indicate the character of these failures. Only recently have we become able to suggest the possibility of race improvement by scientific methods, and only very recently has the possibility appeared in the light of a necessity, the alternative being the universal reward of the unsuccessful.

The present eugenic movement may be said to date from 1865 when Francis Galton showed that mental qualities are inherited just as are physical qualities, and pointed out that this opened the way to an improvement of the race in all respects. The data in support of this pregnant conclusion were included in Galton's work on "Hereditary Genius" published in 1869, when he again emphasized definitely the possibility and desirability of improving the natural qualities of the human race. His suggestions fell upon the stony ground of ignorance even of the most elementary facts of heredity. The subject was raised again in his "Inquiries into the Human Faculty" in 1883, and the word "Eugenics" was then coined. The ground was still non-receptive.

Then followed a period of rapid increase in our knowledge of heredity in animals and plants and in 1901 Galton returned again to the subject, this time in a more direct and elaborate way, and his Huxley Lecture of that year before the Anthropological Institute was upon "The Possible Improvement of the Human Breed under the Existing Conditions of Law and Sentiment." This time he received a real hearing, partly on account of recent disclosures regarding the state of human society and its trends in Great Britain, chiefly because there was at last a real scientific basis for such a proposal. In this lecture, after declaring that the possibility of human race culture is no longer to be considered an academical or impractical problem, Galton proceeded to show that we have a sufficient biological knowledge of man to furnish a working basis. We know of man's variability and heredity—that some men are worth more than others in the community, and that individual traits are also family possessions. This he followed up with definite suggestions as to possible means of the "augmentation of favored stock."

The then recently organized Sociological Society of London took up the subject enthusiastically, and in 1904 and 1905 Galton was invited to deliver addresses before the Society upon this topic. In his first address he spoke upon "Eugenics: its Definition, Scope, and Aims." This proved to be a statement of the elementary principles of the subject—a sort of eugenic creed. Here Galton struck fire. The reading of his paper was followed by very extended discussion and criticism, and he received some enthusiastic support. A few of these enthusiastic supporters brought forth, on the spur of the moment, wonderful, visionary schemes for eugenic progress; much of the adverse criticism went wide of the mark; and, on the whole, Galton must have felt that at least he had demonstrated fully one need for which he had spoken, that of developing a race of able thinkers. Galton's second address before the same society the year following was partly directed at some of this hasty criticism and partly devoted to the setting forth of the possibly ultimate place of the ideals of race improvement in the conscience of the community, and to showing how the whole subject is fraught with "the greatest spiritual dignity and the utmost social importance."

The subject was now fairly launched. Magazine articles appeared on "The New National Patriotism," "Breeding Better Men," *et cetera*. Meanwhile the bio-sociologist settled down to work. And during the five years that have since passed an immense amount of knowledge has been gained, and a large number of excellent workers recruited. Interest in the subject is now general, and its importance recognized as vital. Karl Pearson, known as a good fighter, is Galton's "beak and claws," performing for him much the same kind of service that Huxley performed for Darwin nearly fifty years ago. Galton himself has established a Eugenics Laboratory under the direction of Professor Pearson in the Biometric Laboratory of the University of London and has endowed a Research Fellowship and Research Scholarships. This laboratory is publishing a series of Memoirs and a series of Lectures upon eugenic topics. The University of London is publishing, with the assistance of the Drapers' Company, a series of "Studies in National Deterioration." A periodical, *The Eugenics Review*, is established and appearing regularly. A Eugenics Education Society has been founded to popularize and disseminate the technical information contained in the memoirs and special papers. England remains the seat of greatest activity and interest, but much is being done now in this country. In America the subject is largely under the auspices of the American Breeders Association, which has organized an extremely efficient Committee on Eugenics with which a large number of biological and medical workers are coöperating. This committee has coöperated in the establishment of a Eugenics Record Office, at Cold Spring Harbor, under the direction of H. H. Laughlin. Relevant facts are beginning to pour in from many directions; eugenic ideals are being given practical expression, and the science is rapidly gaining headway.

It may be asked: "Well, what is it all about; are we as a nation not doing well—well enough?" Is it not true, as some have suggested, that this eugenic movement is but one more expression of England's temporary national hysteria transferred to this country? In answer to such queries let us state some of the conditions which have suggested to so many sober thinkers and observers that the time is arriving, has in fact arrived, when we must begin to think of the future of our communities and nations and of our race, rather than contentedly to read of and meditate upon the great achievements of our past, or to parade with self-satisfied air through our glass houses of Anglo-Saxon supremacy. Even were we unthreatened, were we amply holding our own, the mere fact of the possibility of a natural increase of human capacity would make it a practical subject of the utmost importance. We may be sure that somewhere a nation will avail itself of such a possibility as the increase of inherent native talent, physical, mental, moral, and will tend to become a strong and dominant people. Why should not *we* be that people?

It seems that the facts that lead us to think of the future in this matter are of two quite distinct classes. First, we have a great mass of data relative to the composition of our societies and to the changing character of our population, social data of deep significance when broadly viewed and thoughtfully considered. Second, there are certain biological considerations, which all apart from existing social conditions should warn us to be on the lookout. First let us review briefly some of the latter, some of those biological considerations which lead us to regard thoughtfully the problem of the future evolution of man and his societies.

As with other species of animals, each of us comes into the world equipped with a physical constitution and a few simple fundamental instincts. But unlike all other animals, the possession of these alone does not enable us to take and maintain our positions in the community life. Man's life to-day is subject to a great social heritage which, unlike his natural heritage, can be realized only as a result of his own activity and acquisition. Civilized man is the result of Nature plus Nurture. Civilization has been defined as "the sum of human contrivances which enable human beings to advance independently of heredity." The knowledge of fact, historic and scientific, of literature, of art, of custom, and manner, and all that goes to make up the culture and education which are the distinctive traits of our human lives— all this is no possession of ours when we make our first bow to society. Nor do these things become ours through a simple process of growth and development while we remain the passive subjects. All of these things represent the active individual acquirement of the racial accumulation of tradition and learning—what the biologist would call the results of modification. Our troubles begin when we realize that in the acquisition of this load each generation does not begin where the preceding left off, not at all—but we begin where our parents did. The first thing we do toward advancing our places in the world is to absorb what we can of the same kind of thing our forbears absorbed, learn over again their lessons, repeat their experiences; and then we proceed straightway to increase the difficulties for the next generation by writing more books, discovering more facts, making a little more history, and so it goes: the load of tradition increases with every successive generation, and so it has gone since the beginning of man's civilization. It is declared that the modern schoolboy knows more than did Aristotle. We cannot resist the inquiry, Has the modern schoolboy better native ability than had Aristotle? Here is the whole point of this matter; are we any better endowed mentally now that the amount to be mentally absorbed and accomplished is so many times greater? Has our capacity for mental accumulation kept pace with the amount to be accumulated, and with the necessity for such accumulation as a fitting for human life of the civilized variety?

Madison Bentley has recently put it nicely in this way. Does talent grow with knowledge? "May we not suppose that the men and women of some distant glacial age, who dwelt upon the ice, wore the skin of the seal, and ate raw fish, had as much brain and as generous a measure of talent as have their remote descendants who wear sealskins, and eat ices and caviar?" He continues that we have little or nothing to show that the hereditary or innate growth of the mind has kept pace with the growing social heritage; that as regards mental endowment we begin where our distant ancestors began. The chief difference between us and them is that we proceed at once to burden ourselves with information and obligation which for them did not exist. To compass our languages, sciences, histories, arts, the complicated social, political, moral régime, we are supplied with virtually the same minds that primitive man used for his primitive needs. Is it any wonder, he asks, that "education" is the central problem for our or any other advanced civilization?

The biologist asks whether it is not high time to look beyond this artificial bolster of education, to the possibility of actual improvement of the innate mental abilities of man. The student of heredity and evolution looking at this problem has two contributions to make. First, if the mental capabilities of the present race are too limited, increase them; if our minds are too weak to carry the burdens which now must be carried, do not give up the task— strengthen the racial mind. Second, if we should seem to be in danger of developing a stock which is well fitted and able to carry the load of mental acquirement and to push on intellectually, but which is at the same time physically deficient, weak, or sterile, or susceptible to disease, do not let the intellectual capabilities diminish, but build up the physical constitution to a higher supporting level. These are not idle suggestions nor whimsical schemes. The biologist makes them knowing that these things are possible; not only possible, they must be accomplished. We are foolishly building our civilization in the form of an inverted pyramid of individually acquired characteristics. This structure can be made stable only by supplying a broader basis of innate ability which can safely carry the load. This is the first biological warning to sociology.

The second warning we may put in the form in which Ray Lankester in his "Kingdom of Man" has recently presented it so strikingly and which we may abstract freely and with some interpolation. "In Nature's struggle for existence, death ... is the fate of the vanquished, while the only reward to the victors ... is the permission to reproduce their kind—to carry on by heredity to another generation, the specific qualities by which they triumphed." The *origin* of man, partly, at any rate, by such a process of natural selection, is one chapter in his history. Another begins with the development of his mental qualities, which are of such unprecedented power in Nature. These qualities so dominate all else in his "living" activities that they largely cut him off from the general operations of natural selection. Perhaps the only direction in which natural selection is the chiefly operative factor in human evolution to-day is in the development of immunity from infectious disease. Just as man is a new departure in the unfolding scheme of the world, so his presence and characteristics lead to new methods of evolution, of survival, and the like. Knowledge, reason, self-consciousness, will, are new processes in Nature, and it is these which have largely determined the direction of man's history. Nature's discipline of death is more or less successfully resisted by the will of man. Man is Nature's Rebel. "Where Nature says 'Die'! Man says 'I will live.'" By his wits and his will man has overcome many of Nature's bounds and difficulties without changing, as other organisms would, his innate characteristics. Not only this but man has obtained control of his surroundings and at every step of his development he has receded farther from the rule of Nature. Now "he has advanced so far and become so unfitted to the earlier rule, that to suppose that Man can 'return to Nature' is as unreasonable as to suppose that an adult animal can return to its mother's womb."

But at present man puts into operation no real substitute for natural selection. "The standard raised by the rebel man is not that of fitness to the conditions proffered by extra-human Nature, but is one of ideal comfort, prosperity, and conscious joy of life—imposed by the will of man and involving a control, and in important respects a subversion, of what were Nature's methods of dealing with life before she had produced her insurgent son." Progress in the control of Nature has been going on with enormous rapidity during the last two centuries particularly—the "nature searchers" have placed almost limitless power in the hands of men. And yet the builders of society and governments and nations have failed to profit by this increase in natural knowledge. In our social and national organization we remain fixed in the old paths of ignorance. Lankester says: "I speak for those who would urge the conscious and deliberate assumption of his kingdom by Man—not as a matter of markets and of increased opportunity for the cosmopolitan dealers in finance—but as an absolute duty, the fulfillment of Man's destiny." The purpose of his essay is "to point out that civilized man has proceeded so far in his interference with extra-human Nature, has produced for himself and for the living organisms associated with him such a special state of things, by his rebellion against natural selection and his defiance of pre-human dispositions, that he must either go on and acquire firmer control of the conditions, or perish miserably by the vengeance certain to fall on the half-hearted meddler in great affairs." Man is a fighting rebel who at every forward step lays himself open to the liabilities of greater penalties should his attack prove unsuccessful. Moreover, while emancipating himself from the destructive and progressive methods of Nature, man has accumulated a new series of dangers and difficulties with which he must incessantly contend and which he must finally control. Man has taken a tremendous step—created desperate conditions by the exercise of his will—further control is essential in order that he should escape from final misery and destruction.

Nor is this idle, academic invective. The biologist knows that this is true. It is not idle, for man has the means at his command—it is merely a question of their employment. This, then, is the second biological warning to sociology and to statecraft.

Now we may return to consider briefly the nature of those social data which we suggested force us to think seriously of the problem of man's future.

As a primary datum we may note the increasing population of the countries of Europe and North America (Fig. 1). The countries whose population is increasing most rapidly are the United States, Russia, and the German Empire. We know that one important factor of the increase in this country is that of immigration, but this is not sufficient to account for the total. There is continued multiplication of the native population, and of the immigrant after he is here. We wish only to point out in connection with this diagram the steady trend of the population upward, and the fact that obviously somewhere there must be a limit. This cannot go on without end.

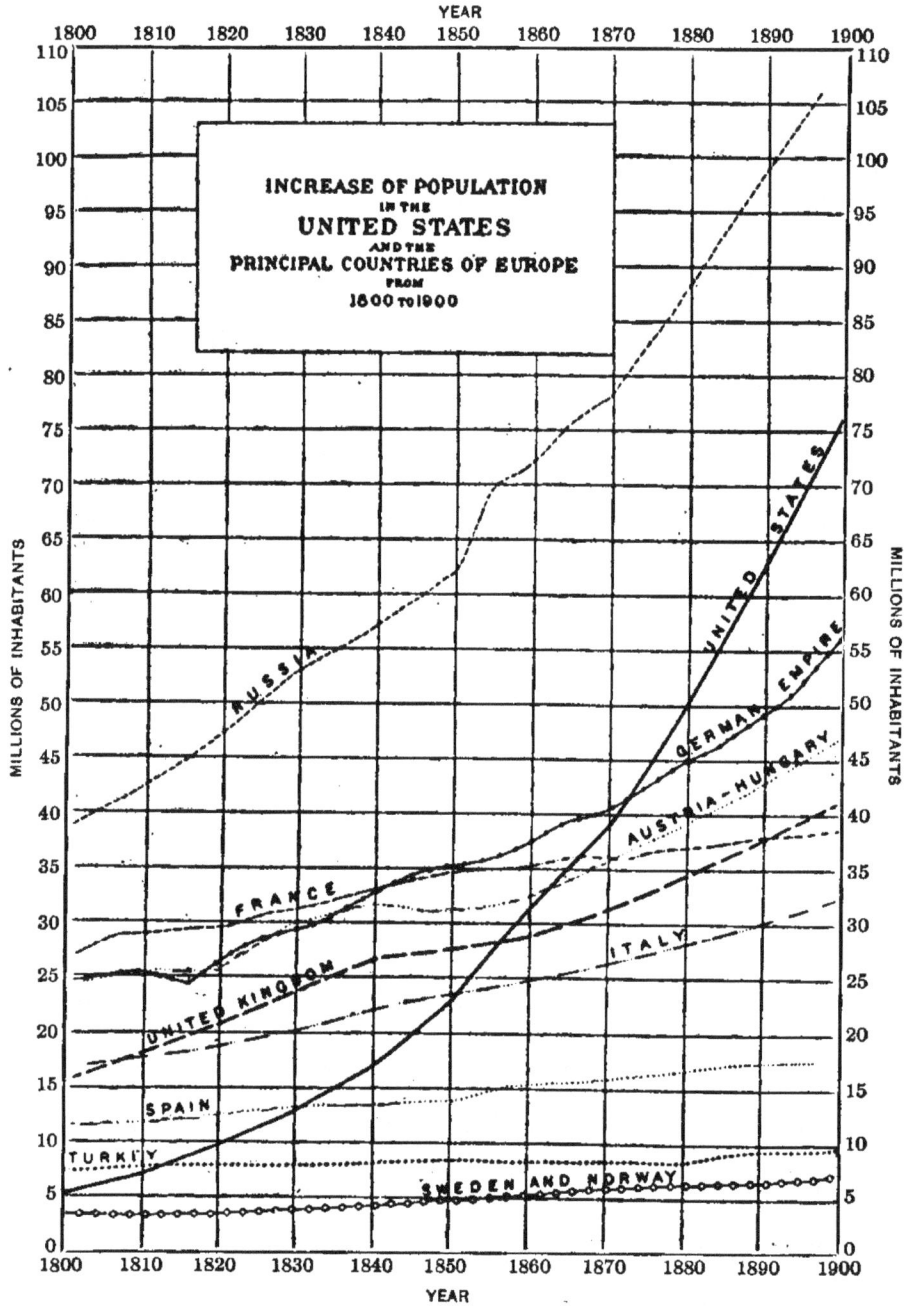

(From "Statistical Atlas," Twelfth Census of the United
States.)
FIG. 1. INCREASE OF POPULATION IN THE
UNITED STATES AND THE PRINCIPAL
COUNTRIES OF EUROPE FROM 1600 TO 1900

An extremely pertinent fact here has been disclosed by Pearson and is based upon very extensive observations among several different classes and nations. It is this—that one fourth of the married population of the present generation produce one half of the next generation. The death rate and the ratio of unmarried to married being what they are, this relation may be stated in this way—twelve per cent of all the individuals born in the last generation produced one half of the present generation. "This is not only a general law, but it is practically true for each class in the community." This conclusion is based upon data from the English, Danish, and Welsh peoples of professional, domestic, commercial, industrial, and pastoral classes, and the per cent of married persons found to be producing one half of each generation varies from twenty-three to twenty-seven with an average of twenty-five per cent. We must ask at once—what is the source of this fourth which is contributing double its quota to the next generation? Is this twenty-five per cent drawn proportionately from all classes of society or are some groups contributing relatively more than others? Is there any relation between this superfertility and the possession of desirable or undesirable characteristics? We may answer at once—there is a distinct and positive relation between civic undesirability and high fertility. We shall return to this subject at the close of the next chapter; only the bare fact is to be mentioned at this time.

It is a matter of common notice and remark that to-day, in England at any rate, there is a dearth of youthful ability. It exists in commerce, science, literature, politics, the bar, the church. We cannot dismiss as merely fashionable the statements that the able classes are not replacing themselves, that men of ability are less able than formerly. Whether or not this is also the condition in America to-day, we know that it soon will be the condition unless steps are taken to bring about a positive relation between civic desirability and ability and the numerical production of offspring.

Let us turn to data of a somewhat different kind. The United States Census Reports for the decades from 1850 to 1900 (1904) include data relative to the number of prisoners in this country. The returns for 1904 omitted certain classes previously enumerated so that for comparative purposes the figures given have to be corrected. On the corrected basis these reports show that the total number of prisoners in the United States increased from 6,737 in 1850 to about 100,000 in 1904, while the total population increased during the same time only from twenty-three to eighty millions (Fig. 2). The ratio of prisoners to the total population is of course the significant relation here, and this increased from 29 per 100,000 in 1850 to 125 per 100,000 in 1904.

Not all of this increase can be attributed to more rigid enforcement of the law or raised standards of morality; there is some reason for thinking that whatever change there has been in these respects has tended to have the opposite effect. We should note, in considering such data as these, that the penologist generally assumes that of the total number of offenders, actually only about ten per cent are in prison at any one time.

During the last century, in France, many parts of Germany, and in Spain the increase in criminality was terrifying. In the United States the number of murders and homicides per million of the entire population has nearly trebled in the last fifteen years (Fig. 2). The average for the five years from 1885 to 1889 inclusive was 38.5 per million, and for the five years from 1902 to 1906 it became 110 per million.

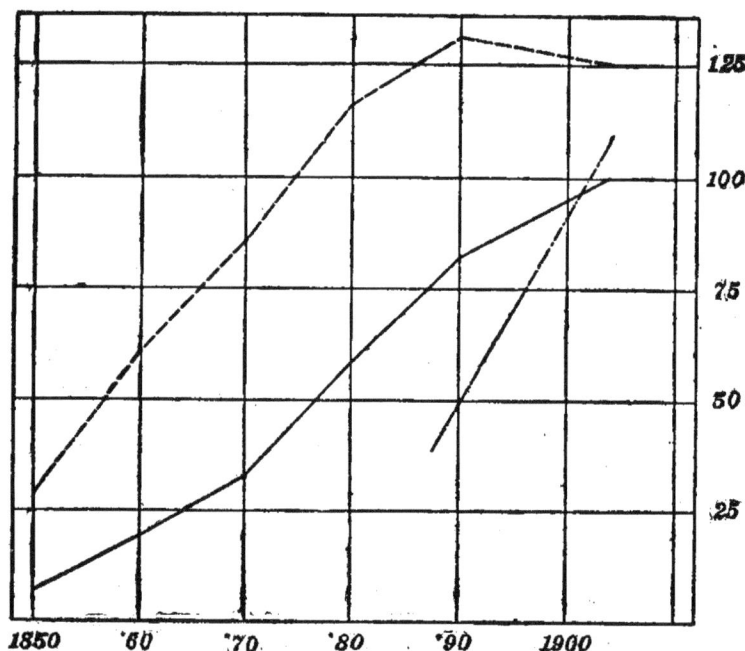

FIG. 2.—Relative and absolute numbers of prisoners in the United States from 1850 to 1904.

------- Number of prisoners per 100,000 of total population.
——— Total number of prisoners (figures to the right are to be read as thousands here).
--·--·-- Number of murders and homicides per million of the total population.

FIG. 2.—Relative and absolute numbers of prisoners in the United States from 1850 to 1904.

England's "defective" classes during the 22 years between 1874 and 1896 increased from 5.4 to 11.6 per thousand of the total; that is, more than doubled in that brief period. Rentoul has collected careful information regarding the number of insane or mentally defective and degenerate in Great Britain. In England the number of "officially certified" insane, which is far less than the actual number, increased from one to every 319 of the total population, to one to 285, in the nine years preceding 1905. In Ireland comparison of the years 1851 and 1896—a period of 45 years intervening—shows an increase in the corresponding ratio from 1:657 to 1:178. The census of 1901 showed in Great Britain 484,507 mental defectives of all kinds; this is one to 85 of the total population, and probably if the whole truth were known the ratio would approximate 1:50, according to Rentoul's calculation. The ratio of known insane just doubled in the decade preceding 1901. The Scottish Commission reports an increase in insane of 190 per cent since 1858, the total population increasing meanwhile by only 52 per cent.

The worst side of these British statistics follows. In 1901, of the 60,000 and more, idiots, imbeciles, and feeble-minded, nearly 19,000—roughly one third—were married and free to multiply; and as for that matter a great many of those unmarried are known to have been prolific. In 1901, of the 117,000 lunatics, nearly 47,000—considerably more than one third—were married. 65,700 idiots and lunatics legally multiplying their kind and worse! Rentoul rightly says: "The hand that wrecks the cradle wrecks the nation."

In the United States the census of 1880 reported 40,942 insane in hospitals, and 51,017 not in hospitals—a total of 91,959 known insane. In 1903 the number in hospitals had increased to 150,151. The number not in hospitals was not given and cannot be determined accurately, but it is conservatively estimated as certainly not less than 30,000, and probably it is far greater than this. In many states it is known that about one fourth of the insane are not in hospitals. But taking the total of 180,000 as a conservative figure, the ratio of known insane in the total population was 225 per 100,000 in 1903 as compared with 183 per 100,000 in 1880.

The methods of the collection of such data vary in different countries so that the results are not comparable. In a single country there is less, though still some, lack of uniformity, so that the exact rate of increase in the ratio of the insane is still somewhat doubtful. Moreover, it is doubtless true that some of this apparent increase results from improved methods in the collection of data, and from more complete registration of these defectives. But suppose we disregard entirely the idea of an increase in the ratio of these defectives, the bare fact of the existence of nearly 200,000 insane in this country is sufficiently alarming; and it is disgraceful to any nation, because it is unnecessary. The Superintendent of the Ohio Institution for the Feeble Minded wrote in 1902: "Unless preventive measures against the progressive increase of the defective classes are adopted, such a calamity as the gradual eclipse, slow decay and final disintegration of our present form of society and government is not only possible, but probable."

The latest census reports for the United States give data relative to the dependents and defectives in institutions. The numbers not in institutions can only be guessed at. But from the available sources we can gain an approximate conception of the numbers in our country to-day as follows:—insane and feeble minded, at least 200,000; blind, 100,000; deaf, and deaf and dumb, 100,000; paupers in institutions, 80,000, two thirds of whom have children, and are also physically or mentally deficient, and to say that one half of the whole number of paupers are in institutions is to give a ridiculously low estimate; prisoners, 100,000, and several hundred thousand more that should be prisoners; juvenile delinquents, 23,000 in institutions; the number cared for by hospitals, dispensaries, "homes" of various kinds, in the year 1904 was in excess of 2,000,000. From these figures we get a rough total of nearly 3,000,000. Must we define a civilized and enlightened nation as one in which only one person in every thirty can be classed as defective or dependent?

It is needless to continue descriptions of this kind. The foregoing are representative data; they are published by the volume. It is always the same story—rapid increase of the unfit, defective, insane, criminal; slow increase, even decrease of the fit, normal, or gifted stocks. It is with such conditions in mind that Whetham writes: "Although this suppression of the best blood of the country is a new disease in modern Europe, it is an old story in the history of nations and has been the prelude to the ruin of states and the decline and fall of empires."

The ultimate aim of Sociology is doubtless the working out of the laws according to which stable communities are formed and maintained, and in which each component individual may enjoy and contribute the maximum of pleasure and profit. So the primary purpose of Statecraft is to produce a nation which shall be stable and enduring. This is all familiar ground. The objects of the nation's immediate activities and concern, protection from enemy, development of commerce and manufacture, agriculture, and education, all these are for the real purpose of establishing and promoting national integrity. No nation exists long without ideals and traditions, without teachers, artists, poets, and yet the primary condition of the existence of all these is a great body of citizens characterized by physical and mental soundness—vigor and sanity. In searching for guiding principles in their great endeavors the sociologist and statesman have sought aid from many sources. But, as Pearson points out, Philosophy has thus far given no law by the aid of which we can understand how a nation becomes physically and mentally vigorous. Anthropology has done little to show wherein exists human fitness as a social organism. Political Economists object that they are not listened to with respectful consideration in legislative chambers. History is the favorite hunting ground of the statesman searching for guidance; but unfortunately history teaches chiefly by example and analogy, rarely by true explanation. And just as some gifted persons are able to give an apt Biblical quotation touching any occurrence whatever, so, many statesmen can cite some historical analogue which they offer as evidence for their views, whatever they are. These men are sincere, in their ignorance of the nature of scientific proof. Finally, although the Statesman still holds rather aloof, the Sociologist comes now to the Biologist, inquiring whether by any chance he may be in possession of data or guiding principles which may be somehow of service in the building of stable societies. The Biologist does not send him away without contribution. The Sociologist makes known his needs, the Biologist displays his possessions, and it is at once evident to both that they have much in common, and that each is able to supply the other with some needed wares. Each may learn from the other; and best of all, the Biologist seems to have information which can be of the greatest service in their common work of building sound societies.

And the biologist is grateful to the sociologist for reminding him that he, too, has sacred duties in this direction. He is too often forgetful that the real aim of his own, as of any science, is to be useful in real human life. It is pleasing to the biologist to feel that he is at last in possession of facts of value to the student of human society, for to him his debt is great. From the sociologist he has drawn the inspirations which have led to some of his greatest discoveries. It was Malthus who suggested to Darwin the great principle of the struggle for existence among men which Darwin so successfully applied to other organisms, and used so profitably in building up his great theory of natural selection. It was from the sociologist that the biologist derived his idea of the physiological division of labor which has proved so fruitful a conception; and from the same source he has drawn many of his conceptions of organic individuality.

We might suggest here some of the topics upon which biology has information of value in this bio-social field; many of these we shall discuss later on from our present and special point of view. First of all come the facts regarding the variability and variation of human beings, not alone in physical characteristics, but in respect to psychic traits as well. Here as in all organisms we must distinguish between true variations and bodily modifications; that is, we must be careful to make, as far as possible, the biological distinction between innate and acquired traits, particularly in considering mental characteristics. Next must come consideration of the facts of heredity. This is undoubtedly the field of greatest importance to the Eugenist; facts of no other kind are of equal significance in determining the course of eugenic practice. We now have a fairly extensive working basis here from which to discuss heredity in man. The various phases of human selection should be noticed, in particular that known as selective fertility or differential fertility in different social groups or classes. Another evolutionary factor of importance here is that of "isolation" in the many and varied forms which it assumes in human society, especially those which result from assortative and preferential mating, and from the operation of social convention, restrictions in marriage, and the like.

Before discussing any of these subjects let us offer here just a word of caution to the enthusiast. The results gained in one field of science cannot be transferred *in toto* to another field and there be found to fit. Biology has learned much from Physics and Chemistry, but the biological applications of the laws of these sciences must be carried out with the greatest care. Such transference has often been premature and attended by results retardative to progress in the field of Biology. Any formula borrowed from one science and applied in another must be rigorously tested under the new conditions. The indiscriminating application of biological laws in the field of sociology may result in confusion and retardation in the progress of both sciences, or at any rate in their practical applications. As Thomson points out in writing on this topic, human society is not only a complex of individual activities of a strictly biological character, but also and further it involves an integration and regulation of those activities which are not yet, at least, susceptible of concrete biological analysis. Thomson says: "The biological ideal of a healthful, self-sustaining, evolving human breed is as fundamental as the social ideal of a harmoniously integrated society is supreme." The great danger here lies in forgetting the fundamental and general character of the biological principles. The ideals of biology and sociology need not coincide, often they do not, but they must not conflict. In practice Eugenics must be largely a social matter; but in its theory, its fundamentals, it must be largely biological.

The coming together of biology and sociology, and their common search for guiding principles in their common endeavor is likely to have results of several kinds. It is likely to bring out more clearly than has yet been done the distinction, in human life and society, between that which is fundamentally biological or animal, and that which is distinctly social. Such information will prove of especial value later when the time comes for the suggestion and carrying out of a definite eugenic program, when the time comes for the real eugenic organization of society. And further the close *rapprochement* of the two subjects will doubtless result in mutual aid and suggestion in the development of each subject in its own stricter field, outside the limits of their common meeting ground.

Before bringing this introductory chapter to a conclusion we should suggest one further caution which must be borne in mind. There may at times seem to be suggestions of antagonism between the biological and the social conceptions of what is eugenic and what is not. Much of this apparent discord will disappear if we recognize that after all the overlapping areas of the two subjects which have fused into the subject of Eugenics are relatively small portions of either whole subject. Sociology has for one of its aims, perhaps its chief aim, the improvement of the present condition of society. The sociologist is interested in the improvement of social conditions to-day and to-morrow. He wants to improve housing conditions, food and milk supplies, to reduce the curses of alcoholism, poverty, and crime, to take the children out of the factory and their mothers out of the sweatshop and put them into schools or under humane conditions of labor. And so on through a long list. The biologist or Eugenist is of course heartily with the sociologist in these endeavors, but as a human being, not as a biologist or Eugenist. For the Eugenist is, as such, by deliberate assumption and definition, directly interested in only such conditions as affect the innate characteristics of the race, conditions which may not have direct reference to the present generation at all, but to the next and to future generations. As a Eugenist he is not concerned with factory legislation, alcoholism, or play grounds, unless it can be shown that there is a relation between these things and the innate mental and physical properties of the race. If there is such a relation, of improvement or impairment, these are eugenic topics; if there is no such relation they are purely social topics, and the Eugenist does not deal with them, not because they are not worth dealing with, but because they are then by definition outside his field. In the end the Eugenist hopes, with the Sociologist, to accomplish these social betterments, but he believes that these will come as by-products in the process of innate racial improvement—improvement in the inherent, physical, mental, and moral qualities of the human kind, and that accomplished in this way the results will be more stable and permanent than any accomplished by attacking the problems as such and separately, largely leaving out of account the real and fundamental cause—bad human protoplasm.

Eugenics is not offered as a universal cure for social ills: no single cure exists. But the Eugenist believes that no other single factor in determining social conditions and practices approaches in importance that of racial structural integrity and sanity. The Eugenist would oppose only those social activities, if such there be, that conflict with his ideal of genuine, progressive, human evolution. The main question which the Eugenist would raise here is largely that of the economy of effort—whether it were not better by concentrating upon a few activities, known to give permanent results, once for all to end an intolerable social condition, rather than to attempt the Sisyphean task.

In conclusion let us quote a few sentences from Francis Galton. "Charity refers to the individual; Statesmanship to the nation; Eugenics cares for both.... I take Eugenics very seriously, feeling that its principles ought to become one of the dominant motives in a civilized nation, much as if they were one of its religious tenets.... Man is gifted with pity and other kindly feelings; he has also the power of preventing many kinds of suffering. I conceive it to fall well within his province to replace Natural Selection by other processes that are more merciful and not less effective. This is precisely the aim of Eugenics. Its first object is to check the birth rate of the Unfit instead of allowing them to come into being, though doomed in large numbers to perish prematurely. The second object is the improvement of the race by furthering the productivity of the Fit, by early marriages and the healthful rearing of their children. Natural Selection rests upon excessive production and wholesale destruction; Eugenics on bringing no more individuals into the world than can be properly cared for, and those only of the best stock."

II

THE BIOLOGICAL FOUNDATIONS OF EUGENICS

II

THE BIOLOGICAL FOUNDATIONS OF EUGENICS

"The gist of histories and statistics as far back as the records reach, is in you this hour,..."

We must now proceed to consider briefly and with only the necessary detail the modes of application of certain biological principles and data in this special field of Eugenics. First of all a clear understanding of the basic ideas of variability and heredity must be had as a primary condition of an appreciation of their significance for the subject before us.

Like any other organism a human being is a bundle of characteristics, physical and psychical. Each person has a definite stature and span, possesses fingers and toes, a head, eyes, ears, hair of a certain color, and so on through a long list of physical traits. Physiological characteristics has he also, such as muscular strength, resistance to fatigue or to disease of many kinds, digestive and assimilative powers, a rate of heart beat, a blood pressure, an habitual gait, posture, a characteristic way of clasping the hands or of twirling the thumbs—and so almost *ad infinitum*. He also possesses certain physiological traits more closely related with the action of the central nervous system—keenness of vision, or hearing, or smell, memory, vivacity, cheerfulness, self-assertiveness, self-consciousness, reasoning power, determination, and the like.

There is a period during the existence of each human being when he does not seem to possess these traits or anything resembling them. For at the beginning of his existence as a new and separate creature, every individual, among the groups of higher organisms, has the form of a single organic cell —the germ. This germ may be, as it is in man, of microscopic dimensions, and it always shows a comparatively slight degree of differentiation of structure. Moreover, the parts and organs of the germ bear no actual or visible resemblance at all to the organs and parts of the organism into which the germ rapidly develops. In other words, in the germ of an organism we have a structure, partly material, partly dynamic, the components of which in some way represent the adult characteristics without resembling them. During the period of the development of the individual, that is to say, during its "ontogeny," these characteristics of the germ become expressed in their final or adult form.

For our purpose it is not necessary to inquire precisely how it is that the structure of the germ can thus represent or determine the structures growing out of it. It must suffice to see that somehow the characteristics of the germ lead to the formation or development of other characters, and these in turn to still others until at last a period of comparative changelessness is reached, when we say that development is completed. It is important to recognize, however, that this development is fundamentally a process of reaction, the reaction between the germ and its surrounding conditions. The characteristics of the adult organism are *determined* primarily by the structure of the germ; they *appear* gradually and successively, as the growing organism reacts to its environing conditions.

An adult organism is continually doing certain things—performing certain movements, producing certain secretions, undergoing a great variety of physical and chemical changes. Just what the organism does at any given moment is in reality determined by two groups of factors: first, it depends, obviously, upon the structure of the organism acting, upon the organs it has to act with, and upon the precise condition of these organs and of the whole individual; and second, it depends upon the nature of those conditions outside of and affecting the organism which lead it to act at all. Either group of factors taken alone will not lead to any activity; activity of an organism must be a reaction between organismal structure and environing conditions —an irritable substance and stimuli to activity. And the character or quality of an act is affected by circumstances within either set of factors.

In much the same way the germ acts, and its action is similarly a reaction between the structure of the germ and its environing conditions. The germ reacts by producing certain parts, differentiating certain structures, in short, by developing. The normal activities or reactions of the adult organism we call in general its "behavior." The normal activities or reactions of the germ and embryo we call "development"; the normal behavior of the germ is development. And in the latter, as well as in the former, changes in either set of factors lead to changes in the nature of the result of their interaction, i. e., to changes in the characteristics actually appearing as the result of development.

In their fully developed state some of the traits or characteristics of organisms are single, simple, fundamental characters, not analyzable into more elementary factors. Such are the number of fingers, or of joints in the fingers, absence of pigments of several kinds from the eyes or hair, presence of cataract, *et cetera*. These so-called "unit characters" are roughly analogous to the chemical elements which may, as units, be combined and recombined in diverse ways, but which always maintain their integrity as elements although different combinations produce wholes that are unlike. Each unit character in the adult is the result of a series of reactions between the environing conditions of development and a germinal structural unit, as yet hypothetical and provisionally called the "determiner," which in some way not yet understood represents this adult trait.

On the other hand, there are many of these things which we call characteristics which seem to be composite, capable of being analyzed or factored into a group of simpler components or unit characters. Such apparently are stature, span, resistance to fatigue, and probably most psychic traits. Each of these complexes results apparently from a series of reactions between the conditions of development and a group of hypothetical germinal determiners that tend to be associated within the germ.

The presence or absence of a determiner in a germ is thus the primary cause of the corresponding presence or absence of a certain characteristic in the adult organism.

But whatever the essential nature of the characteristic in this respect, whether simple or complex, we know further that every organismal characteristic is subject to variation. In any group of human individuals, for example, we can find persons of different stature, different weight, with fingers of different length and form, with heads of different size and shape, hair and eyes of different shades, different blood pressures, pulse rates, digestive possibilities, different degrees of determination, cheerfulness, alertness, and so forth. This fact of variation is not limited to the comparison of the individuals of a given group or generation among themselves, but successive generations considered as the units of comparison show the same sort of thing. And further successive broods from the same parents exhibit this same phenomenon of variation when compared with one another. Variation is a universal fact—not only among organic things but in the inorganic world as well. The variation which any company of persons shows in stature is paralleled by the variation in the diameter of the grains in a handful of sand, or of the drops in a rainstorm.

When we examine the phenomena of variation carefully we find that they are of two quite distinct categories. The first kind of variation, that which we most frequently think of as "variation," should properly be termed *variability*. Differences of this type are small *fluctuations* in any and every character, centering about an average or mean, which is itself fairly definite and fixed—less subject to variation in different groups or through successive generations. For example, if we measure by inches the stature of a thousand or more persons chosen at random we find that they may vary from fifty-four to seventy-six inches; the most frequent heights might be about sixty-nine and sixty-four inches among the men and women respectively. The results of such a measurement may be expressed graphically as in Figure 3, which is an expression of the measurement of 1,052 mothers. The measurement of almost any characteristic in a large group of any organisms usually gives a result of the kind figured. The most significant fact here is that this normal variability exhibited by the traits of living organisms follows closely the laws of chance or probability. That is to say, the number of individuals occurring in any class which has a certain deviation above or below the average, is directly related to, or dependent upon (in mathematical terms, "is a function of"), the extent of the deviation of the value of that class from the average of the whole group. The significance of this is that the precise fluctuation which we find in any individual is the result of the operation of a large number of causes or factors, each contributing slightly and variably to the total result.

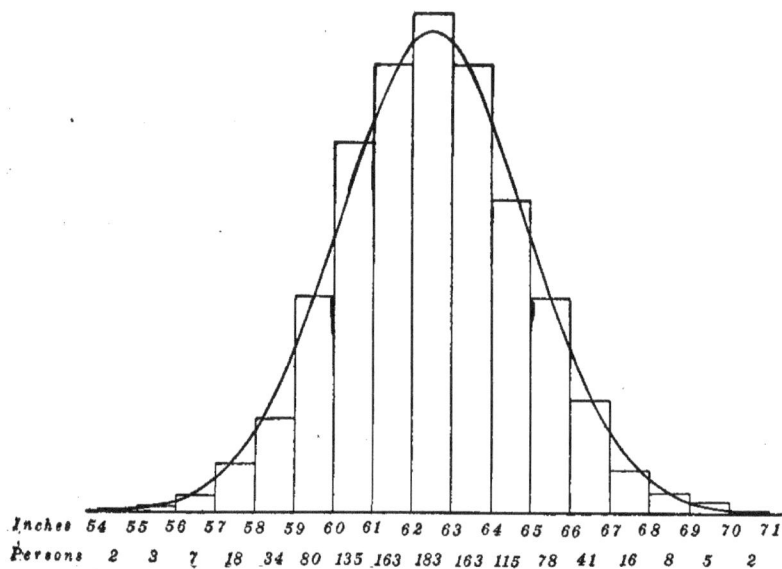

Inches	54	55	56	57	58	59	60	61	62	63	64	65	66	67	68	69	70	71
Persons		2	3	7	18	34	80	135	163	183	163	115	78	41	16	8	5	2

FIG. 3.—Recorded measurements of the stature of 1,052 mothers. The height of each rectangle is proportional to the number of individuals of each given height. The curve connecting the tops of the rectangles is the normal frequency curve. The most frequent height is between 62 and 63 inches. Average height—62.5 inches. Standard deviation, 2.39 inches. Coefficient of variability, 3.8 (2.39=3.8+ % of 62.5 inches). (From Pearson.)

Many of the most important facts about variability can be illustrated by a simple model such as that suggested by Galton. This is a modification of the familiar bagatelle board, covered with glass and arranged as shown in Fig. 4. A funnel-shaped container at the top of the board is filled with peas or similar objects (Fig. 4, A). Below this is a regular series of obstacles symmetrically arranged, and below these, at the bottom of the board, is a row of vertical compartments also arranged symmetrically with reference to the chief axis of the whole system. If we allow the peas to escape from the bottom of the container and to fall among the obstacles into the compartments below we find that their distribution there follows certain laws capable of precise mathematical description, so that it might be predicted with fair accuracy (Fig. 4, B). The middle compartment will receive the most; the compartments next the middle somewhat fewer; those farther from the middle still fewer; and the end compartments fewest. If we connect the top of each column of peas by a curved line we get just such a curve as that given by the stature measurements above (Fig. 3), i. e., the normal frequency curve. A curve of the same essential character would result from plotting the dimensions of a thousand cobblestones, the deviations from the bull's-eye in a target-shooting contest, or by plotting the variability of any organismal character—whether it be the stature or strength of men, the spread of sparrows' wings, the number of rays on scallop shells, or of ray-flowers of daisies.

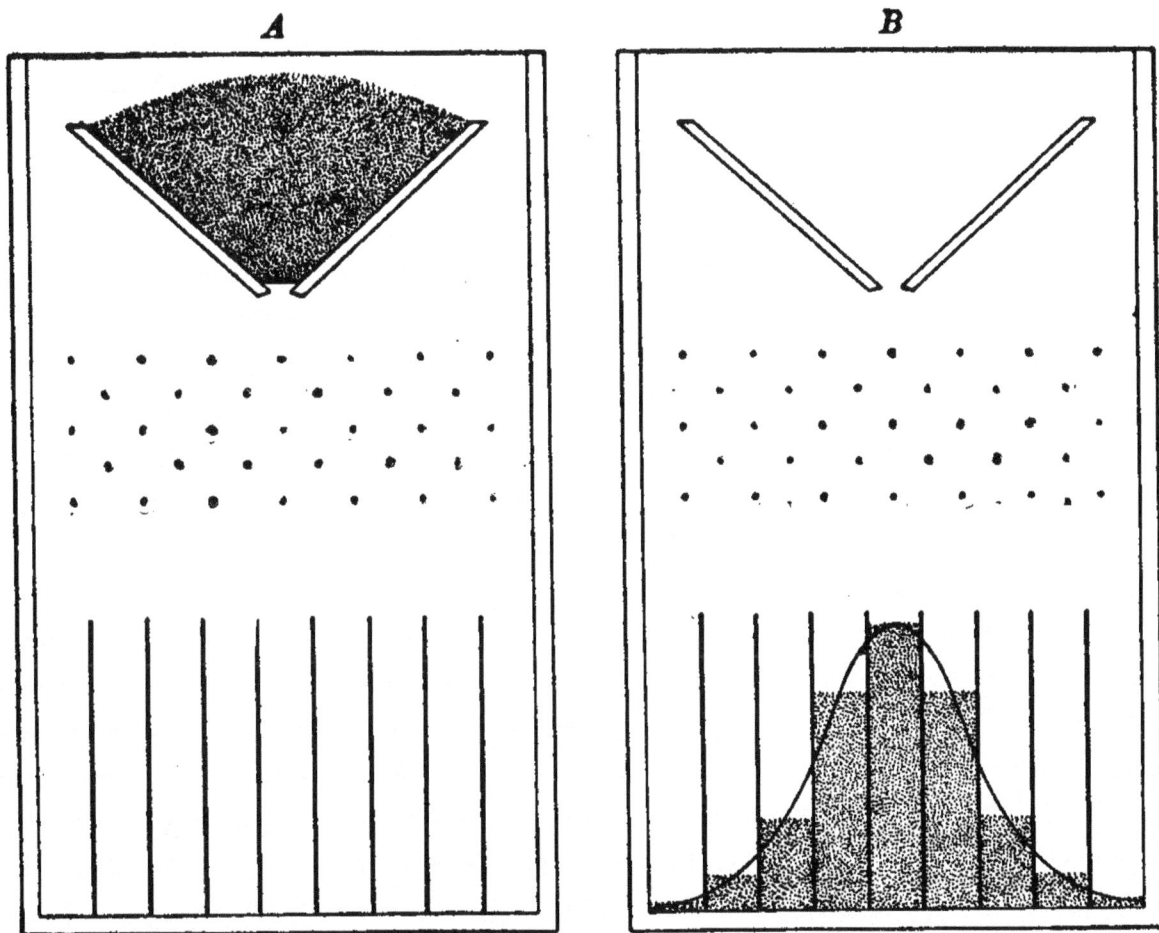

Fig. 4.—Model to illustrate the law of probability or "chance." Description in the text. *A*, Peas held in container at top of board. *B*, Peas after having fallen through the obstructions into the vertical compartments below. The curve connecting the tops of the columns of peas is the normal probability curve.

With this model we may illustrate many other essential facts about variability which must be borne in mind when approaching the problems of Eugenics. Before we allow the peas to fall we know quite definitely what the general distribution of them all will be, but we do not know at all the future position of any single pea. Of this we can speak only in terms of probability; the chances are very high that it will fall in one of the three middle compartments, very low that it will be in one of the extreme compartments. But the chances are equal, whatever they are, that it will fall above or below the average or middle position. We see then that in any group there are many more individuals near the average, i. e., mediocre, than there are in the classes removed from the average and the farther the remove of a class from the average the smaller the number of individuals in that class. Yet all the individuals belong to the same whole group. This leads to the very important fact that *an individual may belong to a group without representing it fairly*. The average individuals are the most representative. But in order to get a correct idea of the whole group we must know, first, to what *extent* deviations occur in each direction, above and below the group average, and, second, the average *amount* by which each individual of the group deviates from this group average. That is, we must know the amount of variability as well as the extent of the greatest divergence from the average. The best measure of the amount of variability exhibited by any group of objects or organisms is not the simple average or mean of all the individual deviations from the average of the group; it is the square root of the mean squared deviations from the group average. This is called the *index* of variability or "standard deviation." In order to make possible the comparison of the variabilities of characteristics measured in unlike units, such as weight and stature, this index must be converted into an equivalent abstract quantity. This is done by reducing the index of variability to per cents of the group average, giving what is called the *coefficient* of variability. Thus, for example, in stature the index of variability (standard deviation) of certain classes of men is approximately 2.7 inches; that is, in a large group of men the amount of individual variation from the average height of 69 inches amounts to 2.7 inches. This gives an abstract *coefficient* of about 4.0 per cent, for 2.7 equals 3.9 per cent of 69. Similarly the index of variability of the weight of a group of university students has been found to be about 16.5 pounds; the average weight is about 153 pounds, and the coefficient of variability is therefore about 10.8 per cent (16.5 equals 10.78 per cent of 153). Although pounds and inches may not be compared, these two abstract coefficients may be, and we may say that men are more than twice as variable in weight as in stature.

Turning now to variation of the second type we find what are ordinarily called *mutations*, or differences quite properly termed *variations*, in a strict sense, as distinguished from the preceding fluctuations or variability phenomena. Mutations or variations are abrupt changes of the average or type condition to a new condition or value which then becomes a new center of fluctuating variability. The difference between variability and variation may be illustrated through an analogy suggested by Galton (Fig. 5). A polygonal plinth, or better a polyhedron, resting upon one face is easily tipped slightly back and forth, but after slight disturbance it always returns to its first position of stable equilibrium. Each face of the plinth or polyhedron represents an organismal characteristic; these slight backward and forward movements represent fluctuations, always centering about the average condition. An unusually hard push sends the plinth over upon another face in which it has a new position of stability; this represents true variation or mutation. In this new position it is again stable, may again be rocked back and forth showing fluctuations about its new average position.

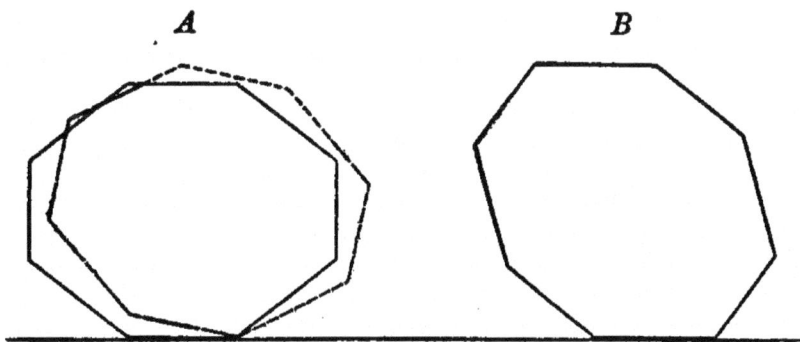

FIG. 5.—Plinth to illustrate the difference between variability (fluctuation) and variation (mutation).

The essential difference between true variation and fluctuation or variability of an extreme nature, is with reference to the inheritance of such divergence. In the second generation the offspring of extreme variates or fluctuations have not the same average as their own parents but an average much nearer that of the whole group to which their parents belonged; the average stature of the children of unusually short or tall parents is respectively greater or less than that of their own parents—that is, is nearer the average of the whole group of parents, provided the shortness or tallness of the parents is a fluctuation. When the shortness or tallness is a true variation or mutational character, offspring have approximately the same average stature as their immediate parents, although the children of course show fluctuation in height so that some are slightly above and others slightly below the parental height.

Mutations may occur through the addition or the subtraction of single characters of the simple or unit type. Such are the variations from brown or blue eyes to albino, five fingers to six, and the like. These are the familiar "sports" of the horticulturalist and breeder. They are of the greatest value in evolution, for it seems quite likely that it is only through the permanent racial fixation of these mutations that permanent changes in the characters of a breed may be effected, i. e., evolution occurs primarily through mutation.

In connection with the general subject of variation we should mention briefly certain aspects of the recent work of Johannsen and Jennings, showing that many organic specific groups or "species," whose characters, when measured accurately give what is called a normal variability curve similar to that of stature illustrated in Fig. 3, are not really homogeneous groups of fluctuating individuals as the curves would indicate superficially, but that each gross group or species is actually composed of a blend of a number of smaller groups, each with its own average and fluctuating variability. It is only when these are taken all together as a lump that they fuse into a single and apparently simple curve.

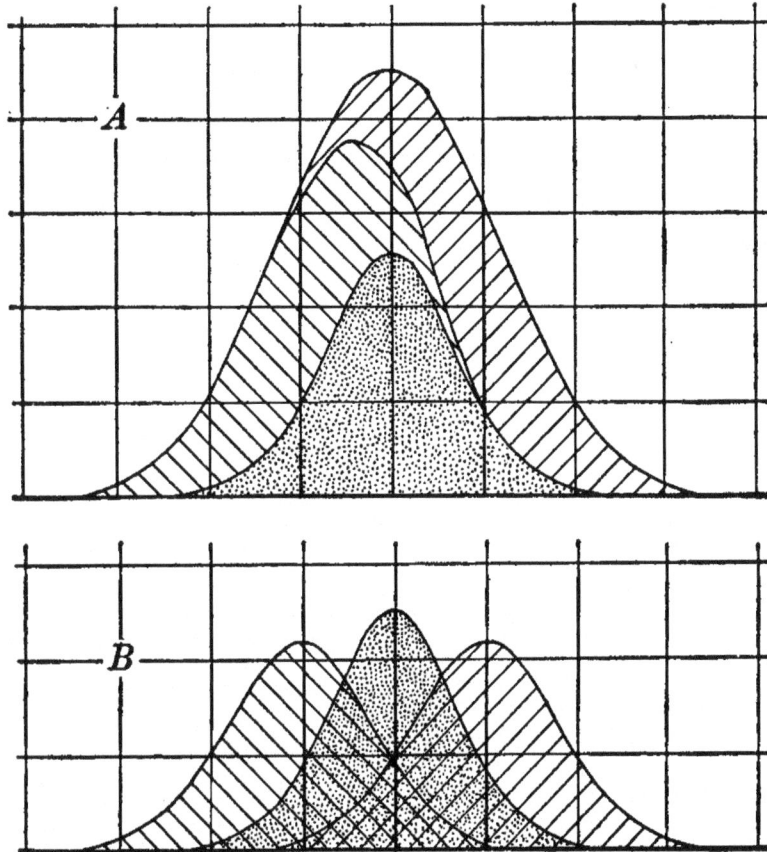

FIG. 6.—Curves illustrating the relation between the pure line and the species or other large group. A, a "species" curve composed of three pure lines. B, the separate elements of the larger curve each with its own average and variability.

For example, the curve shown in Fig. 6, A, which is approximately that of a normal distribution, in some cases might be shown by experimentation to consist in reality of several truly distinct elements, say three for purposes of illustration, as shown in Fig. 6, B. Each of these sub-groups has its own average and its own amount and extent of variability (fluctuation) and it is only by adding them together that we get the larger group. Each of these elementary groups is called a "pure line," which is defined as a group of organisms, all of which are the progeny of a single individual. The characteristics of each pure line remain stable through successive generations, each about its own average; and it is chiefly this fact that enables us to identify the different lines. Transition from the condition of one pure line to another occurs only as a mutation. At present the theory of the pure line is strictly applicable only to organisms reproducing asexually or to self-fertilizing forms where the group observed is actually composed of the progeny of a single organism. It is hardly possible to say as yet whether or not this extremely important theory is essentially applicable to the human species or any species where two organisms are involved in the establishment of a race or line, but there are some indications of a circumstantial nature that it is thus applicable in its essentials and so modified as to include this fact of biparental inheritance.

With this bare skeleton of the subject of variation before us let us see how facts of this kind may have any significance for the subject of Eugenics, any bearing upon the possibility of racial improvement. When any of the varying human traits, and they all vary, is measured carefully and the results tabulated we find that they give us a curve approximating the normal frequency curve, such as we have described above and illustrated in Fig. 3. The coefficients of variability of a great many human traits are known and a few representative coefficients are given in Table I. This type of variability is given then, by measurements of physical characteristics of all kinds, and, what is of greater importance, physiological traits, including mental and moral characteristics, so far as they can be measured by present methods, vary in just the same way. Annual individual earnings give us a curve closely similar to that of a normal frequency curve with an approximate minimum limiting value. Even the tabulation of citizens according to their social standing or "civic worth" gives the same sort of thing. This has been brought out nicely in Galton's discussion of Booth's classification of the population of London.

TABLE I

Coefficients of Variability of Certain Human Traits

Adult Stature	3.6 to 4.0
Length at Birth	5.8 to 6.5
Length of Limb Bones	4.5 to 5.5
Cephalic Index	3.7 to 4.8
Skull Capacity	7.0 to 8.0
Weight (University Students)	10.0 to 11.0
Weight at Birth	14.2 to 15.7

Weight of Brain	7.0 to 10.6
Weight of Heart	17.4 to 20.7
Weight of Liver	14.3 to 22.2
Weight of Kidney	16.8 to 22.5
Lung Capacity	16.6 to 20.4
Squeeze of Hand	13.4 to 21.4
Strength of Pull	15.0 to 22.6
Swiftness of Blow	17.1 to 19.4
Dermal Sensitivity	35.7 to 45.7
Keenness of Eyesight	28.7 to 34.7

It is not so easy to answer the question whether mutations or true variations are occurring frequently in the human species. Usually it is impossible to distinguish between an extreme fluctuation and a true variation without experimental test and the observation of the behavior of the varying trait through several generations. In most instances this has been impossible with human beings. From collateral evidence it seems quite probable that man is mutating with considerable frequency, especially with respect to psychic traits.

The evolution of the race could be directed more easily and permanent results attained more rapidly through taking advantage of valuable mutations than in any other way. A race truly desiring to progress would foster carefully anything resembling mutation in a favorable direction. As a matter of fact, however, our social custom leads us to look with disfavor upon most youthful traits that seem unusual or out of the ordinary. It would be difficult to devise a system of "education" which could more effectively repress than does our own the development of unusual mental traits. In this connection "abnormal" or "eccentric" may often mean a mutation in a profitable direction, a getting away from the average of mediocrity in the direction of improvement.

It is clear that we have the raw materials for race improvement. There are some individuals with more and some with less than the average in any respect—physical, mental, moral. The average of a whole social group can be shifted by subtraction at one end or addition at the other, or more easily and more effectively by both together. In order to raise the general average of the value of any of these traits it is not necessary to strive to exceed the known maximum value in any respect. The study of the "pure line," as mentioned above, shows that this may for a long time remain impossible, or at any rate difficult, pending the appearance of a mutation in a favorable direction. We can, however, raise the general average of physical strength or of mental or moral ability by increasing the relative number of individuals in the upper groups or by diminishing the number in the lower groups, most easily of course and most effectively by doing both of these things. By increasing the numbers composing the lines which form the upper elements of a social group we not only add immensely to the total value of the group but we do actually change somewhat the general average. On the other hand numerical increase in the lines in the lower part of the group will actually lower the average of the whole, though it does not actually affect the number of individuals in the more able and valuable classes.

Another consideration is of great importance here. The average is affected only slightly by the change of individuals from class to class near the average. But the shifting of even one or two per cent of the individuals into or out of extreme positions has a very marked effect upon the character of the total group and upon the average. In the life of the State the character of the general average of the citizens is of the greatest importance, and comparatively small deviations in the average of civic worth may mean much as regards the history of a democracy. Of course the average individuals in a social group may not be those of greatest influence; even when taken all together they may not determine the trend of the life of the society; but that does not alter the essential fact that the condition of the average of the population is of very great moment to a democratic state.

Many of our social endeavors to-day serve in effect to raise individuals from one of the lower groups up to or toward the average. Millions of dollars and an incalculable amount of time and energy are spent annually in striving to accomplish this kind of result. How immeasurably greater would be the benefit to society if the same amount of energy and money were spent in moving individuals from the middle classes on up toward the higher. In the development of our societies we need to use every possible means to carry individuals from positions near the average to positions above the average, and the farther this remove is above the average both in its starting point and its stopping point, the better for the social group. Elevation from mediocrity to superiority has far greater effect upon the social constitution than has elevation from inferiority to mediocrity.

As the Whethams have written recently: "Of late years, the duty of the State to support the falling and fallen has been so much emphasized that its still more important duty to the able and competent has been obscured. Yet it is they who are the real national asset of worth, and it is essential to secure that their action should not be hampered, and their value sterilized, by the jealousy and obstruction of the social failures, and of others whom pity for the failures has blinded. Mankind has been shrewdly divided into those who do things and those who must get out of the way while things are being done, and if the latter class do not recognize their true function in life, they themselves will suffer the most. The incompetent have to be supported partially or wholly by the competent, and, even for their own good, it would be worth while for the incompetent to encourage the freedom of action and the preponderant reproduction of the abler and more successful stocks. It is only where such stocks abound that the nation is able to support and carry along the heavy load of incompetence kept alive by modern civilization."

In discussing the general subject of variation and variability in this connection, we must take always into account the biological distinction between variation and functional modification, between innate and acquired traits. Only the former are of real and primary value in evolution. The distinction is familiar and we cannot dwell upon it here; but it is of particular importance in dealing with social improvement and we shall return to it in the next chapter. Many "social variations" are in reality not variations at all, but modifications; although these may be of the greatest value to the in dividual modified, they are artificial things without permanent value to the race. So many of the distinguishing personal traits are the results of nurture rather than of nature. They represent the result of the incidence of special factors in the environment. It is extremely difficult and at times impossible to distinguish between variations and modifications in adult characters, but in general the distinction is usually clear upon careful analysis.

The changing of the innate characters of the human race is a slow process, depending chiefly upon the advantage taken of the appearance of real mutational variations. On the other hand, it is comparatively easy to improve the condition of the individual by improving his environing conditions—cleaning him, educating him, leading him to higher ideals in his physical and mental and moral life. But as this is easy, so it is impermanent. All this is modificational and has no influence upon the stock. This is not opposed by the Eugenist; it simply is no part of his province, for its effect is not racial. By releasing a deforming pressure it may permit the individual to come back to his real structurally determined condition, but the structural condition itself is not thus affected. It is temporary and must be done over with each generation, or on account of the unfortunate habit of "backsliding," even at intervals shorter than that of a generation.

Let us now turn to another phase of our subject and consider the biological methods of the description and measurement of heredity, as a preliminary to our next chapter in which we shall discuss the bearings of the facts of human heredity upon the possibility of the formation of a permanently improved human breed.

The fact of heredity is one of the most familiar and patent things about organisms. "Do men gather grapes of thorns or figs of thistles?" For we may define heredity as the fact of general resemblance between parent and offspring. This simple definition is disappointing to many persons. "Heredity" is so often supposed popularly to refer only to some occasional, striking, and unusual similarity within a family respecting certain traits or peculiarities. Very often the idea of heredity seems shrouded in mystery: it is some uncanny relation which explains peculiarities and helps the novelist out of difficulties, but is itself inexplicable. In truth, however, the fact that a boy, like his father, has a head and a heart and hands and feet, physical traits characteristic of the human species, that he begins to walk and talk and shave at about the same age as his father did—all this is the fact of heredity. The fact that guinea pigs produce guinea pigs and not rabbits is the fact of heredity. Often it is true that this resemblance is strikingly particular. All know of family traits; we may have our father's eyes or nose, our mother's hair or disposition, a grandfather's determination or a grandmother's patience. But these particular individual resemblances are no more and no less illustrations of heredity than the fact that on the whole children are more like their parents than like other human beings.

The subject of heredity is of supreme importance in the practice of Eugenics. The facts of no other department of biological inquiry are of equal value, and at the same time there is probably no biological subject regarding which there is so much misunderstanding. Of the many phases of this extremely fascinating subject there are chiefly two with which we are particularly concerned as Eugenists. These are the questions: first, how completely are all the distinguishing traits of either or both parents represented in the offspring; and, second, how completely is each trait inherited that is inherited at all? In other words, what we are chiefly interested to know, as bearing upon the subject in hand, is whether all or only some of the characteristics of our parents are heritable, and whether the offspring show each inherited trait with the same intensity shown in the parent, or more, or less.

One of the leading British students of heredity has said that no one should undertake the study of this subject unless he can instantly detect and explain the fallacy involved in the familiar conundrum, "Why do white sheep eat more than black ones?" It is perhaps the elasticity of our language that makes possible the mental confusion involved in this question, but yet it is certainly true that we do tend to confuse individual and statistical statements. We must remember, in connection with this subject particularly, that an individual may belong to a group without representing it, and that within a group there are many more individuals with average than with exceptional characteristics. The mediocre is common, the extremes are rare. And yet an unusual individual may really be an outlying member of a normal group.

In describing the facts of hereditary resemblance between successive generations two formulas are available. One deals ostensibly with the individual—the Mendelian formula: the other deals with the group—the statistical formula. It seems entirely probable that these are not formulas for describing two essentially different processes or forms of heredity, but that in reality these are two ways of describing the same facts seen from two different points of view. The Mendelian formula regards each individual separately and describes its heredity thus. The statistical formula regards the whole group as the unit and considers the individual not as such, but as one of the crowd, concerning which statements can be made only in terms of averages and probabilities; black sheep and white. Of these two formulas the Mendelian is obviously of much the greater importance on account of its more exact, more particular character; its greater definiteness gives it a value in the treatment of eugenic problems that statistical statements must inherently lack. While much has been written of late regarding the Mendelian formula of heredity, we shall find it profitable to repeat here its general outlines and to recall a few of the essential features of this important law that we shall make much use of later.

Let us have a concrete illustration. One of the simplest cases is that of the heredity of color in the Andalusian fowl which has been so clearly described by Bateson. There are two established color varieties of this fowl, one with a great deal of black and one that is white with some black markings or "splashes"; for convenience we may refer to these as the black and white varieties respectively. Each of these breeds true by itself. Black mated with black produce none but black offspring, white mated with white produce none but white offspring. Crossing black and white, however, results in the production of fowls with a sort of grayish color, called "blue" by the fancier, though in reality it is a fine mixture of black and white. At first sight we seem to have a gray hybrid race through the mixture of the black and the white races. Not so: for if we continue to breed successive generations from these blue hybrid fowls we get three differently colored forms. Some will be blue like the parents, some black like one grandparent, some white like the other grandparent. Not only this but we get certain definite proportions among these three classes of descendants. Of the total number of the immediate offspring of the hybrid blues, approximately one half will be blue like the parents, approximately one fourth black, and one fourth white like each of the grandparents. Now comes the most important fact of all. These blacks, bred together produce only blacks, the whites similarly produce only whites; the blues, on the other hand, when bred together produce progeny sorting into the same original classes and in the same proportions as were produced by the blues of the original hybrid generation. Their blacks and whites each breed true, their blues repeat the history of the preceding blues. No race of the hybrid character can be established: blues always produce blacks and whites, as well as blues. A summary of this history in graphic and diagrammatic form is given in Fig. 7.

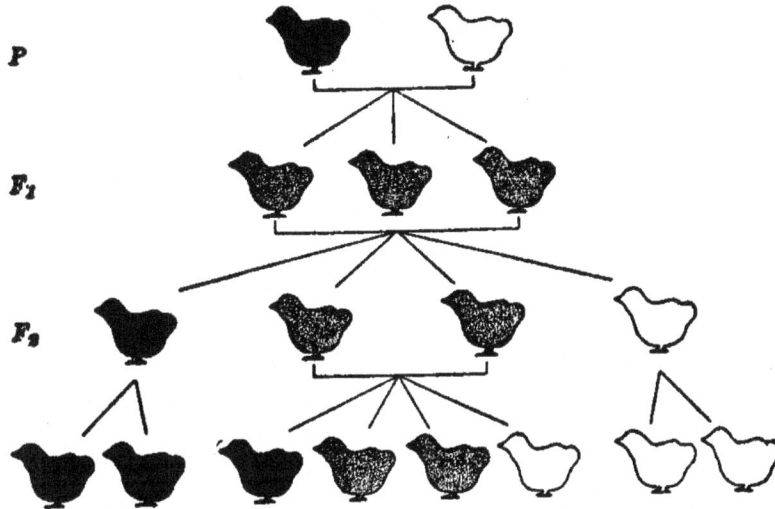

FIG. 7.—Diagram showing the course of color heredity in the Andalusian fowl, in which one color does not completely dominate another. *P*, parental generation. The offspring of this cross constitute F_1, the first filial or hybrid generation. F_2, the second filial generation. Bottom row, third filial generation.

This law of heredity was first discovered about forty-five years ago by Gregor Mendel, working with peas in the garden of the Augustinian monastery in Brünn, Austria. His work curiously failed to arouse the interest of contemporary scientists and his results were soon completely lost sight of. The independent rediscovery of Mendel's formulas of heredity, about ten years ago, was probably the most important event in the history of biology and evolution since the publication of "The Origin of Species."

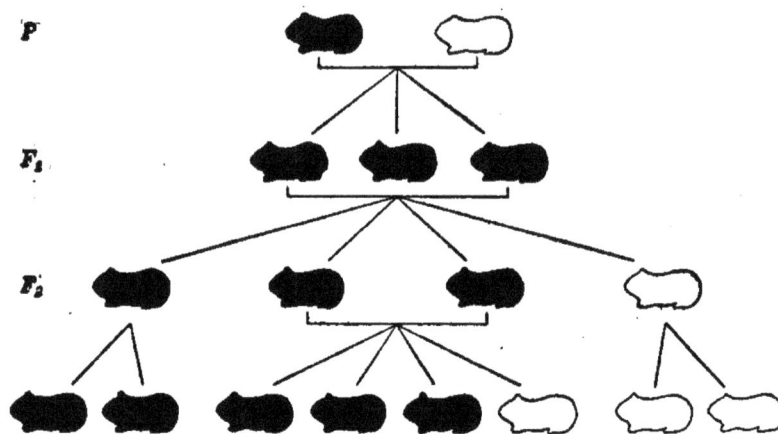

FIG. 8.—Diagram showing the course of color heredity in the guinea pig, in which one color (black) completely dominates another (white). Reference letters as in Fig. 7.

In most cases of Mendelian heredity the progeny are less easily classified than in the case above, because the hybrid individuals resemble one or the other of the parents, quite or very closely. For instance the crossing of the black and white varieties of guinea pigs gives hybrids that are all black like one parent. That is, when the black and white characters are brought together these do not appear to blend into a gray or "blue," as in the case of the Andalusian fowl, but one character alone appears; the black seems to cover up or wipe out the white. This illustrates the frequent phenomenon of *dominance*; one of the two contrasting characters, in this case the black color is said to dominate over the other and the two traits are described as *dominant* and *recessive* respectively. Fig. 8 gives a graphic representation of the history of such a cross. When the black looking hybrids are crossed together the progeny fall into but two groups, one resembling each of the grandparental forms. Three fourths of the progeny now resemble superficially the hybrid form and at the same time one of the grandparents —the dominating black form, while the remaining fourth resembles the other white grandparent. However, we know that the black three fourths do not in reality constitute a homogeneous class but that this includes two distinct groups; one group of one fourth of the whole number of progeny (i. e., one third of all the blacks) are truly black like their black grandparents and in successive generations will, if bred together, produce none but blacks of the same character, i. e., pure blacks: the remaining two fourths of the whole number of progeny (two thirds of all the blacks) in this generation are actually hybrids and in the next generation, if bred together, will give the same proportions of the two colors as were found in the whole of the present generation, i. e., three fourths black, one fourth white. Of these the whites always produce whites, the blacks always produce blacks and whites in the approximate proportions of 3:1; a certain proportion of these—one third (one fourth of the whole generation) always remain blacks, the other two thirds (one half of the whole generation) again produce blacks and whites. In such cases as this where the phenomenon of dominance appears, and this is the usual course of events, it is impossible to say which individuals *are* the hybrids. Only after their progeny are studied can we say which *were* the hybrids.

In the crossing of the black and white Andalusian fowls described above the phenomenon of dominance does not appear; when the two color characters are brought into a single individual neither appears alone, neither overcomes nor is overcome by the other. In the crossing of the black and white guinea pigs dominance is complete; when the two color characters are brought into a single individual only one color appears, the second becomes recessive, that is, it remains present as we know from the later history of such hybrids, but it is not visibly indicated. Besides the Andalusian fowls there are known several other instances of the absence of dominance and there are many cases where dominance is incomplete, i. e., where one character merely tends to dominate the other. And in a few instances dominance is irregular, i. e., sometimes one character dominates, at other times or under other circumstances it does not, as with certain forms of the comb or the feathering of the legs in the common fowl, or with the presence of an extra toe in the domestic cat, the rabbit, and guinea pig. And even in those cases where dominance is said to be complete the trained eye of the breeder can frequently distinguish between the hybrid and the pure bred dominant individuals. The phenomenon of dominance, therefore, is not an essential of the Mendelian theory although it is a frequent, we may say usual, relation.

It does not come within our province to attempt an explanation of this formula of heredity by describing some of the more fundamental conditions upon which it depends. In fact, no complete explanation is yet possible, although several explanatory hypotheses have been suggested. We may outline briefly that which seems the most satisfactory in that it serves to account for most of the facts in Mendelian heredity in a comparatively simple manner. The germ of an organism, we have seen, somehow contains dispositions of materials which primarily determine the characteristics of the organism developed from that germ. To these dispositions or configurations the term of "determiners" has been applied. In a pure variety like the black Andalusians, all the germ cells of each fowl are alike in having this determiner for black color. When two such fowls are mated together their descendants will result from the fusion of two germ cells, *each* containing the determiner for black color; that is, the germ of the new individual comes to have a double determiner, one from each parent, for this trait. In the white variety all the germ cells are alike in *lacking* this determiner; blackness is entirely absent and all their descendants are formed from germ cells entirely without black determiners. When the single germ cell of a black fowl with its single black determiner is fertilized by a germ cell from a white fowl without any determiner for black the resulting hybrid has a color produced by only a single determiner, that from the black parent, and in this case the blackness is not as fully expressed because produced by only this single determiner and the fowl appears gray or "blue"; that is, the black produced by a single determiner is in this case not as black as that produced by the double determiner. Now of course this hybrid fowl forms germ cells containing determiners for color, but these cells, instead of being all alike and with semi-black determiners corresponding with the semi-black characteristics of the individual, are of two different kinds—some are like those of each of the grandparents which fused to give origin to the parent forms, and these are formed in approximately equal numbers—one half with the black determiner, one half without it. When two such fowls are bred together the chances are equal for certain combinations of germ cells; the chances are equal that the "black" or "white" germ cell of the one individual shall meet and conjugate with the "black" or "white" germ cell of the other individual. The result may be expressed algebraically as follows, using the letters B and W to indicate respectively germ cells with and without the black color determiner.

Germ cells of
first parent $\qquad B + W$

Germ cells of second parent		B	+	W	

BB	+	BW		
		BW	+	WW

Combinations in the germ of the offspring					
1BB	+	2BW	+	1WW	

That is, one fourth are pure black (*BB*), one fourth pure white (*WW*), and the remaining half are hybrids, black and white (*BW*). The pure blacks again form germ cells, all possessing the determiner for blackness; the pure whites form germ cells all lacking the determiner for blackness; the hybrid blues produce again equal numbers of germ cells possessing and lacking the determiner for blackness. The relation of the germ cells and the organisms forming them and developing from them is shown in the diagram in Fig. 9.

In the more common cases where the phenomenon of dominance appears, as in the guinea pig, this is explained by saying that here a single determiner for blackness is somehow sufficient to produce the color. In such cases the black color observed may result either from a single (*BW*) or from a double (*BB*) black determiner in the germ which forms the organism. Only when the black determiner is entirely absent (*WW*) does the white color appear in the developed organism and the individual is then said to exhibit the recessive characteristic.

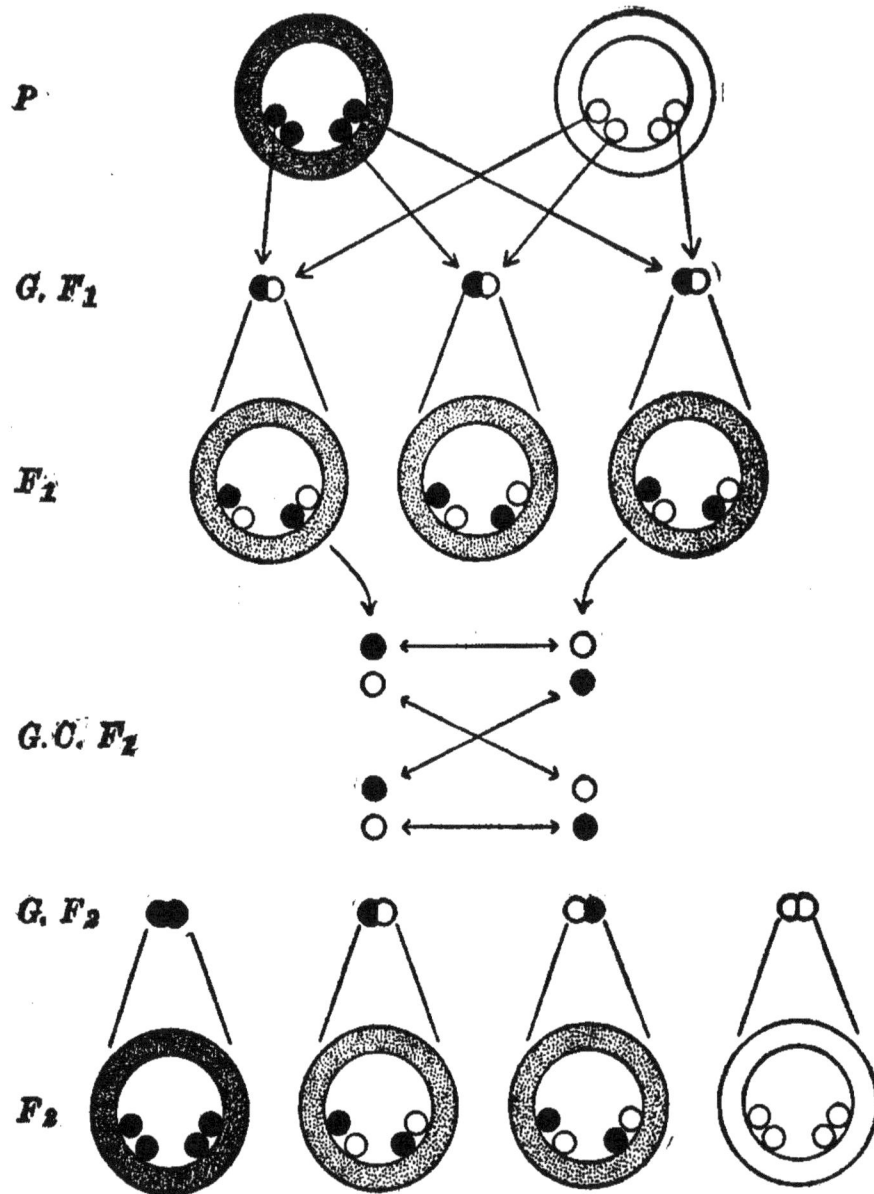

FIG. 9.—Diagram illustrating the relation of the germ cells in a simple case of Mendelian heredity, such as that of color as shown in Figs. 7 and 8. The spaces between the large circles represent the bodies of the individuals while the small circles within each represent the germ cells formed by those individuals. *P*, parental generation; each individual forms a single kind of germ cells. *G. F₁*, germs of the first filial or hybrid generation, each composed of two different kinds of germ cells, one from each parent. *F₁*, individuals of the first filial or hybrid

generation, developed from G. F_1. Each member of this generation forms two kinds of germ cells in approximately equal numbers. G. C. F_1, germ cells of F_1, showing possible combinations resulting from the mating of two members of F_1. Each of these combinations occurs with equal probability. G. F_2, germs of second filial generation resulting from the above random combinations. F_2, individuals of second filial generation. Each now forms germ cells like those which constituted its own germ.

Another possible type of mating is that between a member of a pure race, either dominant or recessive, and a hybrid individual. This form of mating is very common in some of the pedigrees that we shall examine later. The results of such a mating, first between a hybrid and a recessive individual can be most easily described by considering a cross between black and white forms and expressing the result algebraically.

Germ cells of first parent (white or recessive)	W	$+$	W		
Germ cells of second parent (hybrid)	B	$+$	W		
	BW	$+$	BW		
			WW	$+$	WW
	$2BW$	$+$	$2WW$		

That is, returning to the example of the Andalusian fowls, the progeny will be one half hybrid blues and one half whites—no black at all. If the cross had been between black hybrid guinea pigs and white recessive specimens the result would have been half hybrid blacks and half pure whites.

Or supposing the mating to have occurred between the pure dominant (black) and the hybrid the result would have been, in the fowls half pure black and half hybrid blue; in the guinea pig all the progeny would have been black, half pure blacks and half hybrid blacks.

Germ cells of first parent (black or dominant)	B	$+$	B		
Germ cells of second parent (hybrid)	B	$+$	W		
	BB	$+$	BB		
			BW	$+$	BW
	$2BB$	$+$	$2BW$		

In the case of the guinea pigs, although the progeny all look alike (black) their history would show that they were fundamentally unlike, for if crossed with white again the result would be the production of all black looking guinea pigs from the cross with the BB forms, and half black and half white from the BW cross.

On account of the fact of variation every individual is in a certain sense a hybrid. One's two parents have the species characters in common but there are certain distinctive traits that hybridize and follow Mendel's law of heredity. By no means is it to be understood that all individual distinctive traits follow this rule in heredity. Many individual characteristics are what we have learned to call fluctuations—small deviations above or below an average condition of a group. Such differences play no part in Mendelian heredity. Other characteristics may be bodily modifications resulting from the direct reaction between the body tissues and the environing conditions; such traits would not be represented in the organization of the germ cells and consequently would not be inherited at all. At present it seems that the only characteristics that "Mendelize" are those known as "unit characters." Such characters seem to have their origin in real variations or mutations and though each may show fluctuations, these fluctuations in themselves are not hereditary.

This conception of the unit character is an extremely important element in the whole Mendelian theory and it has extended beyond the field of heredity and led to a radical change in our notions of what an organism really is. It is, of course, true in a sense that an organism is a unit, an organism is one thing; but at the same time it is true that an organism is fundamentally a collection of units, of structural and functional characteristics which are really separable things. A few of these units were mentioned in the first pages of this chapter and others are mentioned on a later page. They serve as the building blocks of organisms: individuals of the same species may be made up of similar combinations or of different combinations. One unit or a group of units may be taken out and replaced by others.

From the standpoint of heredity, and particularly from our eugenic point of view, the most important results of the unit composition of the organism lie in the fact that these units remain units throughout successive generations and throughout successive and varying combinations, whatever their associations may be from generation to generation. It is a fact of the greatest eugenic significance that a pure bred individual may be produced by a hybrid mated either with a pure bred or with another hybrid; and that the pure bred resulting will be just as pure bred as any. "Pure bred" now means pure bred with respect to certain traits only. An individual may be pure bred in certain of its characteristics, hybrid in others. Practically there is no such thing as an individual which is either pure bred or hybrid in *all* its traits. One of the chief contributions, then, of Mendelism to the subjects of Heredity and Eugenics is this—that a pure bred may be derived from a hybrid in one generation: the pure bred produced by a long series of hybrid individuals is just as pure as the pure bred which has never had a hybrid in its ancestry. Another important consequent is, that among the offspring of the same parents some individuals may be pure bred and others hybrid.

Community of parentage does not necessarily denote community of characteristics among the offspring. Yet by knowing the ancestry for one or two generations we can know the qualities of the individual. Guesswork is eliminated and the importance of the qualities of the individual is enormously emphasized. It is necessary only to suggest the social and eugenic significance of such facts relating to characteristics that are of social or racial importance.

We shall have occasion in the next chapter to enumerate some of the human unit characters whose heredity has been traced and which have been found to Mendelize, but we may mention here a few Mendelizing units in other organisms in order to give some idea of the kind of character which behaves as a unit and of the range of the forms which have been found to show Mendelian phenomena in their heredity. Among the higher animals one might mention the absence of horns in cattle and sheep; the "waltzing" habit of mice and the pacing gait of the horse; length of hair and smoothness of coat in the rabbit and guinea pig; presence of an extra toe in the cat, guinea pig, rabbit, fowl; length of tail in the cat; and in the common fowl such characters as the shape and size of the comb, presence of a crest or a "muff," a high nostril, rumplessness, feathering of the legs, "frizzling" of the feathers, certain characters of the voice, and a tendency to brood. Among plants may be mentioned such characters as dwarfness in garden peas, sweet peas, and some kinds of beans; smoothness or prickliness of stem in the jimson weed and crowfoot; leaf characters in a great variety of plants; in the cotton plant a half dozen characters have been found to Mendelize; seed characters such as form and amount of starch, sugar, or gluten; flat or hooded standard in the sweet pea; annual or biennial habit in the henbane; susceptibility to a rust disease in wheat. We should not fail to mention that scores of color characters are known to Mendelize, such as hair or coat color and eye color in animals and the colors of flowers, stems, seeds, seed-coats, etc., in plants. The list of Mendelizing traits in different organisms now extends into the hundreds and is increasing almost weekly.

Before leaving the subject of Mendelism we should say that the phenomena, as described above in the Andalusian fowl and guinea pig, are among the simplest known. And while such simple formulas serve to describe the phenomena of heredity in a large number of instances, yet in a great many other cases the descriptive formulas are more complicated. We cannot in this place describe any of these complications. For a full discussion of these and of the whole subject of Mendelism the interested reader is referred to Professor Bateson's work on "Mendel's Principles of Heredity" (1909). It must suffice to say here that in color heredity, for example, such ratios as 9:3:4 or 12:3:1 in the second filial generation instead of the more frequent 1:2:1 or 3:1 are explainable upon essentially the same relations as these simpler and more typical ratios. And further, many less usual Mendelian phenomena, which we cannot undertake to describe here, are associated with what the specialist technically terms "sex limitation," "gametic coupling," and the like.

It is often said that the Mendelian formula has a very limited applicability to human heredity. This is probably true if we consider carefully the grammatical tense in which this statement is made. And yet it is almost certainly true that heredity in man is to be described by this law. This apparent paradox is easily explained. The only characters whose history in heredity follows this formula are the unit characters. A complex trait is not heritable, as a whole, but its components behave in heredity as the separate units. It is perfectly well known that we are deeply ignorant regarding this phase of human structure. Our ignorance here is not the necessary kind, however, it is merely due to the newness of the subject—we have not had time to find out. How can we say that a complex trait is or is not inherited according to some form of Mendel's law when we do not know the nature of the units of which it is composed? We can make no statements about the Mendelian inheritance of such a trait until it is factored into its units. A considerable number of human characteristics are really known to be heritable according to this formula, enough so that several general rules of human heredity have been formulated. But it is also quite within the range of possibility that some traits really do not follow this law, although it cannot yet be said definitely that this is or is not the case. On the whole, then, we cannot, for the next few years, expect too much from the application of Mendel's laws to human heredity, however much this is to be regretted.

Shall we then decline to say anything about the heredity of the great bulk of human characteristics? By no means: we have seen that in our bagatelle board we talk very definitely about the distribution of all the peas, though only about the probable history of one pea. Mendel's law deals with individual inheritance. When we cannot apply this formula we have left still the possibility of talking about human heredity in the group as a whole. That is to say, we have left the opportunity of describing heredity by the statistical methods, with the crowd, not the individual, as the unit. Since we are forced into extensive use of this formula by our present and temporary ignorance of the applicability of Mendel's rule we must get a clear notion of how the statistical method is applied in this matter.

The method is the same as that employed by the statistician in measuring the relatedness of any two series of varying phenomena. If two quantities or characteristics are so related that fluctuations in the one are accompanied in a regular manner by fluctuations in the other, the two quantities or characters are said to be correlated. For instance, the temperature and the rate of growth of sprouting beans are related in such a way that increase in the former is accompanied in a regular way by increase in the latter; or the width and height of the head, or the total stature and the length of the femur similarly vary regularly together so that they are said to be correlated to a certain extent which can be measured. This correlation may result from the fact that one condition is a cause, either direct or indirect, of the other; or there may be no such causal relation between the two phenomena, both resulting more or less independently from a common antecedent condition or cause.

This phenomenon of correlation is not limited among organisms to the comparison of two or more different characters in a single series of individuals; it is applicable also to the comparison of two series of individuals with respect to the same characteristic. Thus we may compare the stature of a series of fathers with the same measurement in their sons. It is this form of correlation with which we are particularly to deal here. While it is not necessary to understand just how this subject is dealt with by the statistician we should know one or two of the elementary principles involved, in order to appreciate the statistical form of many statements about heredity.

The stature of men may be said to vary usually between limits of 62 and 76 inches, the average height being about 69 inches. In the complete absence of heredity in stature we should find that fathers of any given height, say 62 or 63 or 76 inches would have sons of no particular height but of all heights with an average of 69 inches, the same as in the whole group. Or if stature were completely heritable from one generation to the next the *total generations being the units compared,* then 62 or 63 or 76 inch fathers would have respectively sons all 62, 63, and 76 inches tall. When we examine the actual details of the resemblance we find, as a matter of fact, that neither of these possibilities is actually realized. What we do find is that fathers below or above the average height have sons whose average height is also below or above the general average but not so far below or above the general average as were the fathers. If we measured a large number of pairs of fathers and sons with respect to stature we should find each generation with a variability such as that illustrated in Fig. 3 of the stature of mothers, the limits here, however, being about 62 and 76 inches. But if we measured all the sons of 62-inch fathers they would be found to vary say from 62 to only 69 inches, averaging about 66 inches. Similarly 63-inch fathers would have sons from 62 to 70 inches tall, averaging about 66.5 inches, or 76-inch fathers might have sons from 69 to 76 inches in height, averaging about 72 inches, and so on for fathers of all heights. In general, then, we may say that fathers with a characteristic of a certain plus or minus deviation from the average of the whole group have sons who on the whole deviate in the same direction but less widely than the fathers, although the fact of variability comes in so that some few of the sons deviate as widely as, or even more widely than, the fathers, others deviate less widely than the fathers from the average of the whole group. This is the general and very important statistical fact of *regression.*

The phenomenon of regression may be made somewhat clearer by the aid of a simple diagram—Fig. 10. Here are plotted first the heights, by inches, of a group of fathers, giving the series of dots joined by the diagonal *AB.* Next are plotted the average heights of the sons of each class of fathers: 62-inch fathers give 66-inch sons, 63-inch fathers 66.5-inch sons, 64-inch fathers 67-inch sons, and so for all the classes of fathers. These dots are then joined by the line *EF.* This is the *regression line.* Had it been the case that there was no regression in stature the different classes of fathers would have had sons averaging just the same as themselves and the line representing the heights of the sons would have coincided with the line *AB.* Or if regression had been complete the fathers of any class would have had sons averaging about 69 inches—just the same as the average of the whole group—and the line representing their heights would have had the position of *CD* in the diagram. As a matter of fact, however, neither of these possibilities is actually realized and the regression line *EF* is approximated in an actual series of data. A similar relation has been found for many characters other than stature.

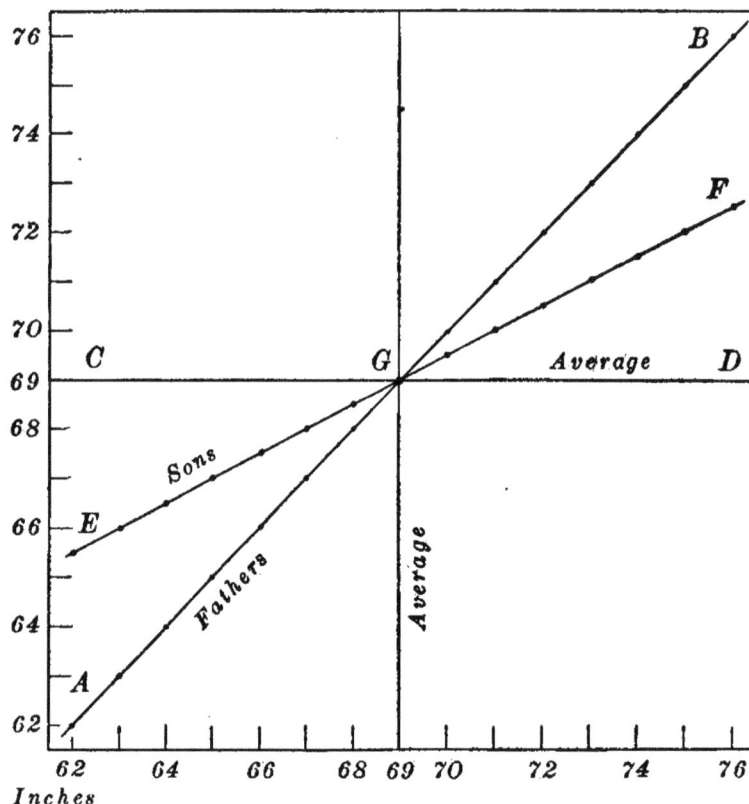

FIG. 10.—Diagram illustrating the phenomenon of regression.
Explanation in text.

The fact of regression is of considerable importance for the theory of evolution as well as for the subject of Eugenics when describing the phenomena of heredity in this statistical manner in whole groups without paying attention to particular individuals. Regression is found in all characteristics observed in this way, psychic as well as purely physical. "The father [i. e., fathers] with a great excess of the character contributes [contribute] sons with an excess, but a less excess of it; the father [fathers] with a great defect of the character contributes [contribute] sons with a defect, but less defect of it."

Now, whatever the actual extent of this regression is in a group we need to know how uniformly it occurs for all the classes of different
   deviations from the general average,
that is, we need to know whether the extreme groups regress to the same relative extent as do those nearer the general average; and, further, we need to know how nearly the sons of fathers of any certain height are grouped about their own average. In other words, we should know, first, whether the regression of the sons of 62 and 76 or 67 and 71 inch fathers is proportionately the same in each case, and, second, to what extent the sons of 62-inch fathers vary, whether they vary as do the fathers of 62-inch sons, and so for each group. This kind of information we get by calculating what is called the *coefficient of heredity*. The calculation of this coefficient is a complicated process which it is unnecessary to describe here. It must suffice to say that a numerical coefficient can readily be determined, which will express the average closeness and regularity of the relationship between all the plus and minus deviations from the group average in fathers and the corresponding plus and minus deviations from the group average of their sons with respect to a given characteristic. This coefficient of heredity may vary between 0.0 and 1.0. When it is 0.0 there is, on the whole, no regularity in the relationship, i. e., no heredity; when it is 1.0 there is, on the whole, complete regularity, i. e., heredity is complete. Neither of these values is ever actually found in determining coefficients of heredity in the parental relation; these are usually between 0.3 and 0.5. It should be emphasized again that this comparison is between whole groups and not between individuals, and that it fails to allow for the distinction between fluctuations and true variations. And, further, it should be noted that the information derived from such a coefficient is defective in that it takes into account only the relationship between the son and one parent; the maternal relation is just as important but this has to be determined separately. There is no satisfactory method of determining the relation between children and both parents at the same time.

The coefficient of heredity is, therefore, an abstract numerical value which gives us a fairly precise estimate as to the probable closeness of the relation between deviations from the group average of any character in two groups of relatives. The coefficient of *correlation* is, in general, a measure of the relation between two different characteristics or conditions in a single group of individuals. The method of its determination and its limiting values are the same as for the coefficient of heredity.

By experience the coefficients of heredity and correlation in general are found to have the following significance:

> 0.00-   no relation.
> 0.00-0.10— no significant relation.
> 0.10-0.25— low; relation slight though appreciable.
> 0.25-0.50— moderate; relation considerable.
> 0.50-0.75— high; relation marked.
> 0.75-0.90— very high; relation very marked.
> 0.90-1.00— nearly complete.
> 1.00— complete relation.

One further point remains to be considered, which applies not so much to coefficients of heredity as to coefficients of correlation in general, i. e., to the relatedness of two different characters or series of events in a single group of cases or individuals. This is that coefficients of correlation may be either positive or negative. That is, the real limits of the value of the coefficient are plus one and minus one. The example given above of stature of fathers and sons gives a positive coefficient. Whenever the deviation from the average of one group is accompanied in the second group by a deviation in the same direction, the coefficient is positive. A negative correlation means that deviation from the average in a given direction in the first group is accompanied in the second group by a deviation in the opposite direction. If we imagine that as one measurement increased above its average a second related measurement decreased below its average the correlation in such a case would be negative. For instance, if we measured the relation between the number of berry pickers employed and the quantity of berries remaining unpicked, in a number of different fields we would get a negative correlation coefficient. Some organisms are formed in such a way that increase in one dimension, such as length, is associated with decrease in another, such as breadth; measurement of the relatedness of these dimensions would give a coefficient of correlation that might be very high, indicating a considerable relation in the deviations, but it would be negative. In an instance of negative correlation the relation is that of "the more the fewer." As we shall see presently, a negative correlation may be just as important and significant as a positive correlation.

The application of the principles of heredity to our subject of Eugenics is of such great importance that it is reserved for separate consideration in the next chapter. We may, therefore, devote the remainder of this chapter to the consideration of data of another kind, which are commonly treated by this same method of determining correlation coefficients between two sets of varying phenomena in order to determine whether there is any actual relation between them or not. This will serve to illustrate the use of this method.

We shall turn then to the subject of differential or selective fertility in human beings and consider its relation to Eugenics. As a starting point we may take the self-evident statement that a group of organisms will tend to maintain constant characteristics through successive generations only when all parts of the group are equally fertile. If exceptional fertility is associated with the presence or absence of any characteristic the number of individuals with or without that trait will either increase or diminish in successive generations, and the character of the distribution of the group as a whole will gradually become altered, the average moving in the direction of the more fertile group. Or if infertility is so associated, then the average of the whole group moves away from that condition. Eugenically, then, we should ask whether in human society there is at present any such association of superfertility or infertility with desirable or undesirable traits. It is obviously the aim of Eugenics to bring about an association of a high degree of fertility with desirable traits and a low degree of fertility with undesirable characteristics.

First, let us look at certain data gathered relative to the size of the family in both normal and pathological stocks (Table II). In order that a stock or family should just maintain its numbers undiminished through successive generations and under average conditions, at least four children should be born to each marriage that has any children at all. The table given shows clearly what stocks are maintaining, what increasing, and what diminishing their numbers.

TABLE II

Fertility in Pathological and Normal Stocks. (From Pearson)

	AUTHORITY.	NATURE OF MARRIAGE. (Reproductive period.)	NO. IN FAMILY.
Deaf-mutes, England	Schuster	Probably complete	6.2
Deaf-mutes, America	Schuster	Probably complete	6.1
Tuberculous stock	Pearson	Probably complete	5.7
Albinotic stock	Pearson	Probably complete	5.9
Insane stock	Heron	Probably complete	6.0
Edinburgh degenerates	Eugenics Lab	Incomplete	6.1
London mentally defective	Eugenics Lab	Incomplete	7.0
Manchester mentally defective	Eugenics Lab	Incomplete	6.3
Criminals	Goring	Completed	6.6
English middle class	Pearson	15 years at least, begun before 35	6.4
Family records— normals	Pearson	Completed	5.3
English intellectual class	Pearson	Completed	4.7
Working class N. S. W.	Powys	Completed	5.3
Danish professional class	Westergaard	15 years at least	5.2
Danish working class	Westergaard	25 years at least	5.3
Edinburgh normal artisan	Eugenics Lab	Incomplete	5.9
London normal artisan	Eugenics Lab	Incomplete	5.1
American graduates	Harvard	Completed	2.0
English intellectuals	Webb	Said to be complete	1.5

All childless marriages are excluded except in the last two cases. Inclusion of such marriages usually reduces the average by 0.5 to 1.0 child.

This subject has been investigated recently in a rather extensive way by David Heron, for the London population. Heron concentrated his attention upon the relation of fertility in man to social status. He used as indices to social status such marks as the relative number of professional men in a community, or the relative number of servants employed, or of lowest type of male laborers, or of pawn-brokers; also the amount of child employment pauperism, overcrowding in the home, tuberculosis, and pauper lunacy. Twenty-seven metropolitan boroughs of London were canvassed on these bases, which are certainly significant, though not infallible, indices to the character of a community. His results are shown in the briefest possible form in Table III.

TABLE III
Correlation of the Birth Rate with Social and Physical Characters of London Population. (From Heron.)

	CORRELATION COEFFICIENT.
With number of males engaged in professions	-.78
With female domestics per 100 females	-.80
With female domestics per 100 families	-.76
With general laborers per 1,000 males	+.52
With pawnbrokers and general dealers per 1,000 males	+.62
With children employed, ages 10 to 14	+.66
With persons living more than two in a room	+.70
With infants under one year dying per 1,000 births	+.50
With deaths from pulmonary tuberculosis per 100,000 inhabitants	+.59
With total number of paupers per 1,000 inhabitants	+.20
With number of lunatic paupers per 1,000	+.34

inhabitants

This table gives the results of the calculation of coefficients of correlation between the birth rates and the conditions enumerated. We may just recall that this coefficient is a measure of the regularity with which the changes in two varying conditions or phenomena are associated: and further that a coefficient of 1.0 indicates perfectly regular association, 0.75 a very high degree of regularity. The first line of the table then, for example, means that when these twenty-seven districts were sorted out, first, with reference to the number of professional men dwelling in them, and then with reference to their respective birth rates, there was found a very high degree of regularity (coefficient of correlation=-.78) in the association of these two conditions—birth rate and number of professional men. Here is a very close relation, *but*, the sign of the coefficient is *negative*. The significance of this negative sign is that among the communities studied those where the number of professional men is the larger show always, at the same time, the lower birth rates. Coming to the second line of the table, it seems fair to assume that the number of servants employed in a district in proportion to the total number of residents or families there, gives a fairly though not wholly satisfactory indication of the social character of the community. Measurement of the actual relation between the proportional number of servants employed in a community and the birth rate in that community, gave practically the same result as in the case of the number of professional men. The more servants employed in a district the lower its birth rate. Two methods of measuring this relation gave essentially the same result; comparison of the birth rate with the ratio of domestics, first to the number of families, second to the number of females, gave -.76 and -.80 respectively—very high coefficients and both negative.

But the sign changes and becomes positive when we come to other comparisons. When we count the relative number of pawnbrokers and general dealers, of "general laborers" (that is, men without a trade and without regularity of occupation and employment), of employed children between the ages of ten and fourteen, of persons living more than two in a room, when we consider the infant death rate, the death rate from pulmonary tuberculosis, and the relative number of paupers,—then we find the signs of the coefficients are all positive, and on the average the coefficients are more than 0.50—a moderate to high degree of regularity of the relation. The districts characterized by the larger numbers of such individuals or by higher death rates of these kinds, are at the same time the districts where the birth rates are the higher.

In a word, then, Heron found that the greater the number of professional men, or of servants employed in a community, the lower the birth rate—a very high degree of negative correlation. On the other hand, the more pawn-brokers, child laborers, pauper lunatics, the more overcrowding and tuberculosis, the higher the birth rate—a high degree of positive correlation. Little doubt here as to which elements of the city are making the greater contributions to the next generation. There may be some doubt, however, so let us consider two possible qualifications of these results. First, is not the death rate also higher among these least desirable classes? Yes, it is. Is it not enough higher to compensate for the difference in the birth rates, so that after all the least desirable classes are not more than replacing themselves? No, it is not. After calculating the effect of the differential death rate among these different social groups it still remains true that the *net* fertility of the undesirables is greater than the *net* fertility of the desirables: the worst classes are in reality more than replacing themselves numerically in such communities; the most valuable classes are not even replacing themselves. Second, is not this the same condition that has always existed in these districts? Why any cause for supposing that this is going to bring new results to this society? Has not such a condition always been present and always been compensated for somehow? Fortunately, Heron is able to compare with these data of 1901 similar data for 1851, and is able to show that every one of these relations has changed in sign since that date—in fifty years. The significance of this change in sign is probably clear. It means here that in London sixty years ago there was a high degree of regularity in the relation such that the more professional men and well-to-do families the community contained, the higher the birth rate; that ten years ago this had all become changed so that the more of these desirable families found in a district the lower is the birth rate. It means that sixty years ago the relation was such that the more undesirables numbered in a district, the lower its birth rate; ten years ago the more undesirables, the higher the birth rate, and the coefficients of 1901 are unusually high, indicating great closeness and regularity in this relation. Heron is further able to show that as regards number of servants employed, professional men, general laborers, and pawnbrokers in a district, the intensity of the relationship has *doubled*, besides changing in sign, in the period observed. It is not necessary to review the history of this change nor to discuss the causes involved, but it is necessary to take into account for the immediate future the fact of the change.

Sidney Webb has recently published an account of the birth-rate investigations undertaken by the Fabian Society with a view to determine the causes leading to the rapidly falling birth rate in England. During the decade previous to 1901 the number of children in London actually diminished by about 5,000, while the total population increased by about 300,000. As far as they bear upon this phase of the subject his results fully confirm these we have been considering. The falling off is chiefly in the upper and middle classes, in the classes of thrift and independence, and it has occurred chiefly during the last fifty years. Webb cannot find that this is due to any physical deterioration in these classes; it is due to a conscious and deliberate limitation of the size of the family for what are thought prudential and economic reasons.

An actual reduction in the number of children may not be an unmixed evil. A falling birth rate may be a good sign. This is partly a question for the political economist. "Suicide" may be a socially fortunate end for some strains. But when, in either a rising or a falling birth rate, we find a differential or selective relation, then the subject is eugenic. If the higher birth rate is among the socially valuable elements of each different class the Eugenist can only approve; to bring about such a relation is one of his aims. What we really find, however, is the undesirable elements increasing with the greatest rapidity, the better elements not even holding their own.

One further aspect of the result of the smaller family remains to be considered. Are the various members of a single family approximately similar in their characteristics or are the earlier born more or less likely to be particularly gifted or particularly liable to disease or abnormal condition? Or is there no rule at all in this matter? There is much evidence that the incidence of pathological defect falls heaviest upon the earlier members of a family. Consider, for example, the presence of tuberculosis. We should ask, in families of two or more, are the tubercular members, if any, as likely to be the second born or third or tenth as to be the first born?

The data are tabulated in Fig. 11, *A*. The distribution of family sizes being what it is in the number of families investigated and tabulated, we should expect that there would be about 65 tubercular first born, 60 tubercular second born, and so forth, on the basis of its average frequency in the whole community, provided the chances are equal that any member of the family should be affected with tuberculosis. What we actually find, however, is that 112 first born are affected, about 80 second born, and after that no relation between order of birth and susceptibility to tuberculosis. That is, susceptibility to tuberculosis is double the normal among first born children. The same thing is true for gross mental defect. Fig. 11, *B*, shows that the ratio of observed to expected insane first born children is about 4 to 3. Such a relation has long been known to criminologists and frequently commented upon. Fig. 11, *C*, gives a definite expression to the facts here. Whereas, in the number of families observed about 56 criminal first born were to be expected, the number actually found is about 120; for the second born the corresponding numbers are about 54 and 78, and after that no marked relation is found between order of birth and criminality. For albinism (Fig. 11, *D*) the expected and observed numbers among first born are about 185 and 265, second born 165 and 190, and thereafter no definite relation. It remains to be seen whether a similar relation holds for the unusually able and valuable members of a family; something has been said on both sides here, but there are available at present no data sufficiently exact to be worthy of consideration.

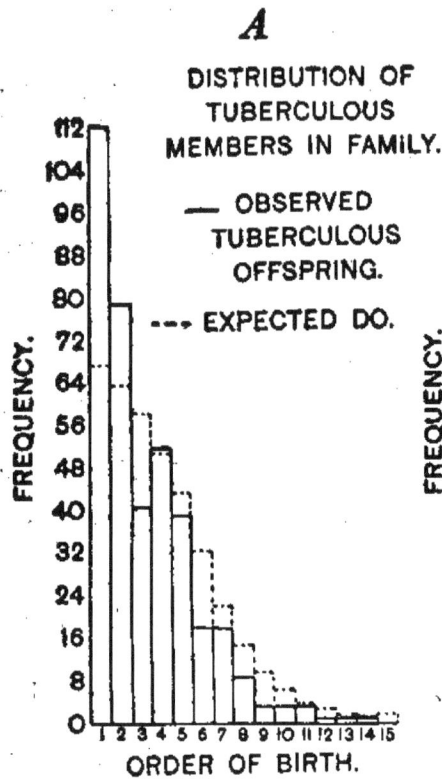

A

DISTRIBUTION OF
TUBERCULOUS
MEMBERS IN FAMILY.

—— OBSERVED
TUBERCULOUS
OFFSPRING.

---- EXPECTED DO.

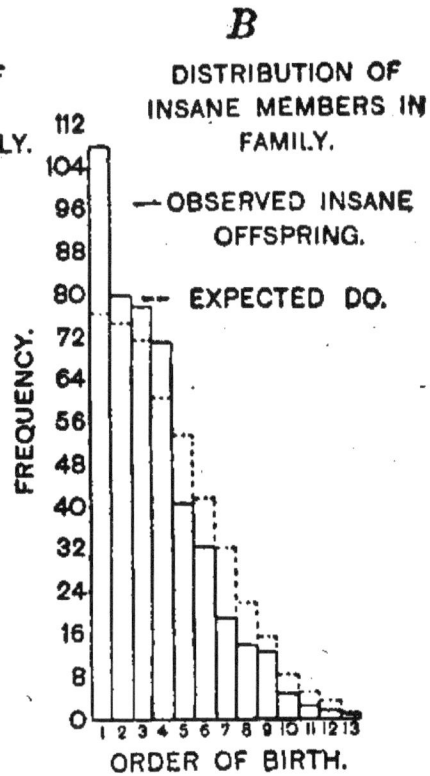

B

DISTRIBUTION OF
INSANE MEMBERS IN
FAMILY.

—— OBSERVED INSANE
OFFSPRING.

---- EXPECTED DO.

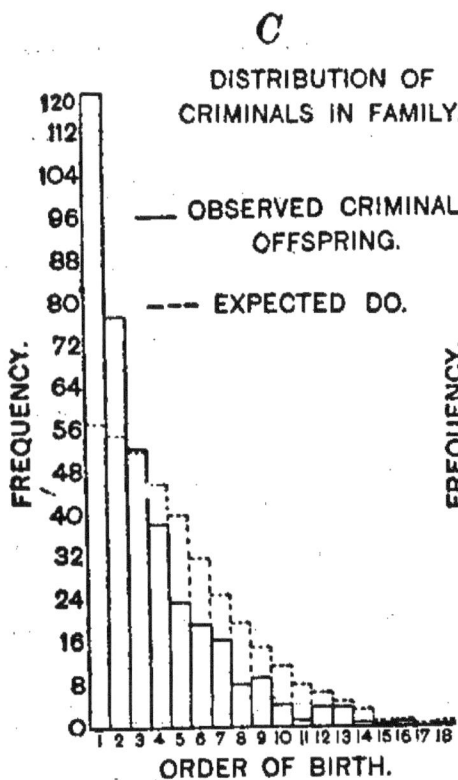

C

DISTRIBUTION OF
CRIMINALS IN FAMILY.

—— OBSERVED CRIMINAL
OFFSPRING.

--- EXPECTED DO.

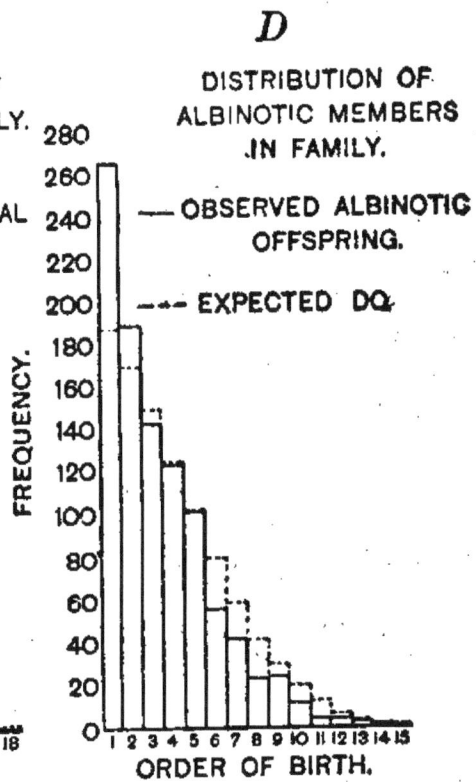

D

DISTRIBUTION OF
ALBINOTIC MEMBERS
IN FAMILY.

—— OBSERVED ALBINOTIC
OFFSPRING.

--- EXPECTED DO.

FIG. 11.—Diagrams showing the relation between order of birth and incidence of pathological defect. (From Pearson).

We have here a result that has very important bearings upon the value to the race of the large family and of the danger of the small family. The small family of one, two, or three children contributes on the average much more than its share of pathological and defective persons. No matter just now what the causes are, they seem to be more or less beyond remedy. The result for the future, however, must be reckoned with. This relation has important bearings upon the custom of primogeniture as well as upon the eugenic values of the large family.

In conclusion let us give a few sentences only slightly modified from Pearson's "Grammar of Science." The subject of differential fertility is not only vitally important for the theory of evolution, but it is crucial for the stability of civilized societies. If the type of maximum fertility is not identical with the type fittest to survive in a given environment, then only intensive selection can keep the community stable. If natural selection be suspended there results a progressive change; the most fertile, whoever they are, tend to multiply at an increasing rate. In our modern societies natural selection has been to some extent suspended; what test have we then of the identity of the most fertile and the most fit? It wants but very few generations to carry the type from the fit to the unfit. The aristocracy of the intellectual and artizan classes are not equally fertile with the mediocre and least valuable portions of those classes and of society as a whole. Hence if the professional and intellectual classes are to be maintained in due proportions they must be recruited from below. This is much more serious than would appear at first sight. The upper middle class is the backbone of a nation, supplying its thinkers, leaders, and organizers. This class is not a mushroom growth, but the result of a long process of selecting the abler and fitter members of society. The middle classes produce relatively to the working classes a vastly greater proportion of ability; *it is not want of education, it is the want of stock which is at the basis of this difference.* A healthy society would have its maximum of fertility in this class and recruit the artizan class from the middle class rather than *vice versa.* But what do we actually find? A growing decrease in the birth rate of the middle and upper classes; a strong movement for restraint of fertility, and limitation of the family, touching only the intellectual classes and the aristocracy of the hand workers! Restraint and limitation may be most social and at the same time most eugenic if they begin in the first place to check the fertility of the unfit; but if they start at the wrong end of society they are worse than useless, they are nationally disastrous in their effects. The dearth of ability at a time of crisis is the worst ill that can happen to a people. Sitting quietly at home, a nation may degenerate and collapse, simply because it has given full play to selective reproduction and not bred from its best. From the standpoint of the patriot, no less than from that of the evolutionist and Eugenist, differential fertility is momentous; we must unreservedly condemn all movements for restraint of fertility which do not discriminate between the fertility of the physically and mentally fit and that of the unfit. Our social instincts have reduced to a minimum the natural elimination of the socially dangerous elements; they must now lead us consciously to provide against the worst effects of differential fertility—a survival of the most fertile, when the most fertile are not the socially fittest.

The subject before us illustrates the direct bearing of science upon moral conduct and upon statecraft. The scientific study of man is not merely a passive intellectual viewing of nature. It teaches us the art of living, of building up stable and dominant nations, and it is of no greater importance for the scientist in his laboratory, than for the statesman in council and the philanthropist in society.

III

HUMAN HEREDITY AND THE EUGENIC PROGRAM

III

HUMAN HEREDITY AND THE EUGENIC PROGRAM

"A breed whose proof is in time and deeds;
What we are, we are—nativity is answer enough to objections."

A few years ago official recognition was taken of the disturbing fact that the annual wheat yield of Great Britain was grossly deficient in both quantity and quality. In 1900 The National Association of British and Irish Millers, with almost unprecedented sagacity, raised a fund to provide for a series of experiments under the direction of a competent biologist, in order to discover if possible some means of restoring the former yield and quality of the native wheats. The story of the result reads like a romance. The experimenter—Prof. R. H. Biffen—collected many different varieties of wheat, native and foreign, each of which had some desirable qualities, and studied their mode of inheritance. Now, after only a few years of experimentation a wheat has been produced and is being grown upon a large scale in which have been united this desirable character of one variety, that character of another. From each variety has been taken some valuable trait, and these have all been combined into one variety possessing the characteristics of a short full head, beardlessness, high gluten content, immunity to the devastating rust, a strong supporting straw, and a high yield per acre. A wheat made to order and fulfilling the "details and specifications" of the growers.

Manitoba and British Columbia opened up whole new lands of the finest wheat-growing capacity, but the season there is too short for the ripening of what were the finest varieties. This new specification was promptly met and the early ripening quality of some inferior variety was transferred to the varieties showing other highly desirable qualities, and these countries are now producing enormous quantities of the finest wheat in the world.

All of this has been made possible by the discovery, mentioned in the preceding chapter, that many characteristics of organisms are units and behave as such in heredity; they can be added to races or subtracted from them almost at will. Pure varieties breeding true can be established permanently by taking into account the Mendelian laws of heredity. Similar results have been accomplished in many other plants and in many animals. A cotton has been produced which combines early growth, by which it escapes the ravages of the boll weevil, with the long fiber of the finest Sea

Island varieties. Corn of almost any desired percentage of sugar or starch, within limits, can be produced to order in a few seasons. The hornless character of certain varieties of cattle can be transferred to any chosen breed. Sheep have been produced combining the excellent mutton qualities of one breed with the hornlessness of another, and with the fine wool qualities of still a third. And so on from canary birds to draft horses. New races can be built up to meet almost any demand, with almost any desired combination of known characters, and these races remain stable. Possibilities in this direction seem to be limited only by our present and rapidly lessening ignorance of the facts of Mendelian heredity in organisms —facts to be had for the looking.

What is man that we should not be mindful of him? Why should we utilize all this new knowledge, all these immense possibilities of control and of creation, only for our pigs and cabbages? In this era of conservation should not our profoundest concern be the conservation of human protoplasm? "The State has no material resources at all comparable with its citizens, and no hope of perpetuity except in the intelligence and integrity of its people." As Saleeby puts it: "There is no wealth but life; and if the inherent quality of life fails, neither battle-ships, nor libraries, nor symphonies, nor Free Trade, nor Tariff Reform, nor anything else will save a nation."

In this work of the creation and establishment of new and valuable varieties, two essential biological facts are made use of. The raw materials are furnished by variation—by the fact that there are individual and racial differences. The means of accomplishing results are furnished by heredity —the fact that offspring resemble the parents, not only in generalities, but even in particulars, and according to certain definite formulas.

And, further, in the formation and establishment of a new race of plant or animal a conscious and ideal process is involved. The will of some organism guides the process, carefully doing away with hit and miss methods, and proceeding as directly as may be possible to an end *desired*. The facts of variation and heredity are sufficiently demonstrated for all organisms other than man; are they true of man also? Have we available the possibilities for the improvement of the human breed? If not, Eugenics is merely an interesting speculation. We have mentioned already the facts of variation in man; we undoubtedly do have the raw materials. What about heredity, and what about the directive agency? Let us look now at some of the facts of human heredity and consider some of the possibilities in the way of directive agencies. Is it going to be possible to breed a stable human race permanently with or without definite characteristics which now appear only in certain groups, or sporadically as variations?

At the outset we should say that the knowledge of human heredity is as yet largely of the statistical sort. We know how a great many characters are inherited, on the average. The subject of Mendelian heredity is so new that there has been hardly time to investigate more than a few human characteristics from this point of view. Certain conditions add to the difficulties here. First, many, probably most, of the more important human traits are complexes, not units, and it is a long and difficult process to analyze them into their units, with which alone Mendelism deals. Second, in human society we cannot carry on definite experiments under controlled conditions, directed toward the solution of some concrete problem in heredity. It is true that Nature herself is making such experiments constantly, but at random, and rarely under ideal conditions of what the experimenter calls control or check. We have first to seek and find them out, and when they are found we often discover that there are lacking many of the facts essential to a complete or satisfactory analysis of the facts displayed. The comparatively small size of the human family sometimes makes it difficult to get data sufficiently extensive to be really significant. And the long period that elapses between successive human generations adds to the difficulty of getting precise information, for in dealing with the heredity of some traits comparisons must be made with individuals of the same ages, and the period of observation of a single observer seldom exceeds the duration of a single generation. Yet in spite of all these difficulties we have a fairly broad and exact knowledge of human heredity in respect to some characteristics.

Human heredity involves both physical and psychical characters—both the body and the mind are concerned. Among other animals little if anything is known regarding psychic inheritance, but the physical traits of men are inherited in just the same ways and to the same degrees as in animals. This degree or intensity of inheritance may be expressed in coefficients of heredity between the groups of relatives being compared. To mention a few examples of coefficients for physical traits we have the following:

CHARACTER OBSERVED	PARENTAL COEFFICIENT	FRATERNAL COEFFICIENT
Stature	.49-.51 }	.51-.55 }
Span	.45 }	.55 }
Fore Arm	.42 } .47	.49 } .53
Eye Color	.55 }	.52 }
Hair Color		.57 ⎯ Average
Hair Curliness		.52
Head Measurements-three		.55 ⎯ "
Cephalic Index (Ratio between breadth and length of cranium)		.49

We might give many others, but it is unnecessary. Notice that these parental and fraternal coefficients group about an average value of about .50 or slightly less. Similar coefficients have been worked out for other degrees of relationship; thus grandparental coefficients are about .25.

Stated briefly, in less exact terms, these coefficients mean that, with respect to such traits as deviate from the group average, the resemblance of brothers and sisters to each other or of children to their parents is, on the whole, approximately mid-way between being complete in its deviation from the average and in not deviating at all from the average in the direction of the fraternal or parental characteristic. Grandchildren tend to deviate from the group average only about one fourth as far as their grandparents. It should be remembered that these are statistical and not individual statements, and that as many "exceptions" will be found in the direction of greater resemblance as in that of lesser resemblance.

One of the present objects of the student of heredity, perhaps his chief object, is to be able to state the facts of human heredity in Mendelian terms, reducing many of the complex human traits to their simpler elements. Some of the chief objections to the use of the statistical formula of heredity are that apparently it is applicable only to the fluctuating variabilities of organisms; that it rarely takes into account the presence of (and therefore the heredity of) true variations or mutations—and we have seen that it is just these characters that are of the greatest value in evolution; and that heredity is after all fundamentally an individual relation which loses much of its definiteness and significance when we merge the individual in with a crowd. To some these seem fatal objections to any use of the statistical formula and it is certainly true that they greatly limit its value. But for the present at least the statistical statement of certain facts of heredity is still useful in this bio-social field. We may therefore use the statistical formulas of heredity as a kind of temporary expedient, enabling us to make statements regarding inheritance of certain characters in the group or class, pending the time when we shall be able to give the facts a more precise and more "final" expression in Mendelian formulas. Many human traits are indeed already known to Mendelize. Most of these are, however, "abnormal" traits or pathological conditions; we are still in the dark regarding the actually Mendelian or non-Mendelian inheritance of most of man's normal characteristics. We might enumerate the following Mendelizing human characters—eye color, color blindness, hair color and curliness, albinism (absence of pigment), brachydactylism (two joints instead of three in fingers and toes), syndactylism (union of certain fingers and toes), polydactylism (one or more additional fingers or toes in each hand or foot), keratosis (unusually thick and horny skin), hæmophilia (lack of clotting property in the blood), nightblindness (ability to see only in strong light—a retinal defect usually), certain forms of deaf mutism and cataract, imbecility, Huntington's chorea (a form of dementia).

In observing Mendelian heredity we should bear in mind that a given character may be due either to the presence or to the absence of a "determiner" in the germ. Long hair such as is characteristic of many "Angora" varieties of the guinea pig and cat, for example, is believed to be due to the absence of a determiner which stops its growth. Blue eyes are due to the absence of a brown pigment determiner, *et cetera*. The presence or absence in the offspring of such characters as we know do Mendelize can be predicted when we know the parental history for two generations.

Turning now to the inheritance of mental traits and including, of course, moral traits here as well, we find that we are almost entirely limited to the statistical statement of results. Pearson found upon examining data from a large number of school children, brothers and sisters, that the coefficients of heredity between them were the same as for their physical traits. His results are summarized in Figure 12. The physical traits measured were, in the order plotted in the figure—health, eye color, hair color, hair curliness, cephalic index (ratio between breadth and length of cranium), head length, head breadth, head height. These gave an average of .54 in brothers, .53 in sisters, and .51 in brothers and sisters. The psychical traits in order were—vivacity, assertiveness, introspection, popularity, conscientiousness, temper, ability, handwriting. The corresponding averages were .52, .51, .52.

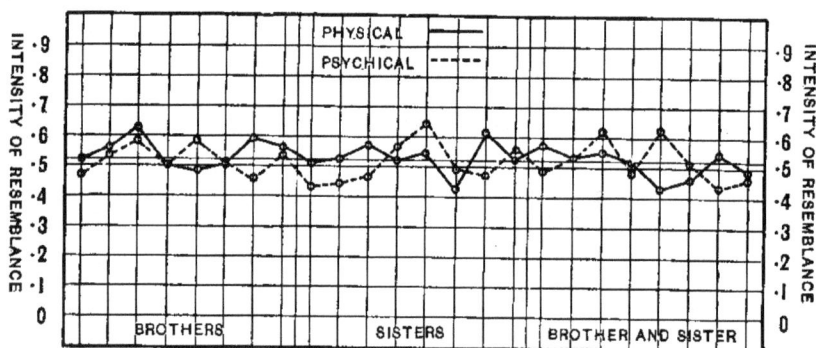

FIG. 12.—Coefficients of heredity of physical and psychical characters in school children. Characters enumerated in text. (From Pearson.)

Galton's pioneer works on "Hereditary Genius," "English Men of Science," and "Natural Inheritance" showed with great clearness the fact of mental and moral heredity. Wood's recent extensive study of "Mental and Moral Heredity in Royalty" shows the same thing, although not all the results of these investigations are given in mathematical form. Little can be said regarding Mendelian heredity of mental traits because the psychologist has not yet told us how to analyze even the common and simpler psychic characters into their fundamental units; since we do not know what the mental hereditary units are, obviously we cannot work with them. Much of our knowledge in this field does not permit of very accurate summary, though pointing indisputably to the fact of mental inheritance in spite of the very great influences of training and education, environment and tradition, in moulding the mental and moral characteristics—influences with much greater effect here than in connection with physical characters.

Galton studied the parentage of 207 Fellows of the Royal Society, a Fellowship which is a real mark of distinction. He assumed that one per cent of the individuals represented by the class from which his observations were drawn, that is the higher intellectual classes, might be expected to be "noteworthy": among the general population the average is really about one in 4,000 or one fortieth of one per cent. On the one per cent basis Galton found that Fellows of the Royal Society had noteworthy fathers with 24 times the frequency to be expected in the absence of heredity; noteworthy brothers with 31 times the expected frequency; noteworthy grandfathers 12 times; and so on through various grades of relationship.

Schuster examined the class lists of Oxford covering a period of 92 years and found that first honor men had 36 per cent first or second honor fathers; second honor men had 32 per cent first or second honor fathers; ordinary degree men 14 per cent first or second honor fathers. These percentages are far in excess of that to be expected—perhaps 0.5 per cent—on the assumption that ability is not inherited. Schuster also determined the coefficients of heredity between fathers and sons as regards intellectual ability, the evidence being class marks in Oxford and Harrow; these he found to be about .3 for the parental relation and .4 for the fraternal. The intensity of heredity in many forms of insanity has been determined and this runs up much higher—.57 parental and .50 fraternal.

It is clear I take it, that the fact of human heredity does not concern only physical traits but extends to psychical traits as well, and with about the same intensity. This fact has been found true also for still less analyzable characters such as length of life, fertility or infertility and the like, and again about the same intensity of resemblance is found.

Human heredity is a fact then just as human variability is a fact. We have truly the raw materials and the means for racial improvement. The ability to direct the evolution of the human race makes this our supremest duty.

The facts of human heredity can more easily be brought home to us by the examination of some actual pedigrees and family histories. We may look at a few representative cases which will serve to bring out some additional aspects of the significance to society of the demonstrated fact of heredity. In the examination of single family histories we should remember that a single pedigree may not accurately illustrate a general law of heredity—again, an individual case may belong to a group of cases without representing them fairly. Even in observing illustrations of Mendel's laws allowance has to be made for the variability due to "chance" meetings of germ cells. It is only when large numbers of individuals are observed that the typical Mendelian fractions and ratios can be strictly observed. It must be borne in mind then that the histories given below illustrate the nature of the facts of heredity rather than the laws of heredity. Some special cautions in the interpretation of certain pedigrees will be suggested in particular cases. Many of the figures are taken from the extremely valuable "Treasury of Human Inheritance," now being published by the Eugenics Laboratory of the University of London. In these figures and some others a uniform series of symbols is used. Successive horizontal lines designated by Roman numerals indicate generations; within a single generation the individuals are numbered consecutively simply for purposes of reference. The meaning of the more common symbols is as shown in Table IV. We may first consider a few pedigrees showing the heredity of physical abnormalities or defects.

HUMAN HEREDITY

Table IV.

Symbols used in Pedigrees. As adopted by the Galton Eugenics Laboratory.

♂ ♀ [♀]   Male and female respectively, not possessing the trait under consideration.

♂ ●   Male and female possessing the trait under consideration.

○   Unknown sex—normal or affected.

⊖   Trait incompletely developed.

⊘   Neither presence nor absence of trait can be affirmed.

⊗   With a deformity or disease of special character which may possibly be associated with that under consideration.

○—○   Twins.

③   Indicates number of children.

♂ ♀   Marriage.

?   Number of children unknown.

   Number and character of children unknown.

S. P.   *Sine prole.* (No offspring.)

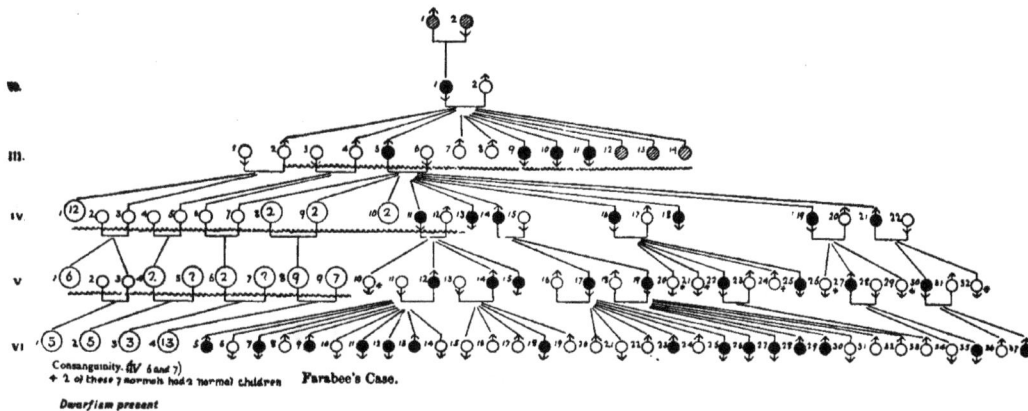

FIG. 13.—Family history showing brachydactylism. Farabee's data. (From "Treasury of Human Inheritance.")

Fig. 13 illustrates a family history where brachydactylism (an abnormality of the digits commonly called shortfingeredness, due to the lack of one joint in each digit) is present and frequently associated with dwarfism. We may describe this case rather fully because it illustrates nicely the heredity of a trait according to the Mendelian formula. The parentage of the affected female (II, 1) who started this line is uncertain. The marriage was with a normal male whose parentage is unknown but evidently normal. This pair produced 11 children, the character of 8 of whom is known; 4 were affected, 4 unaffected, a Mendelian ratio resulting from the mating of a normal with a hybrid individual, the observed character dominating (i. e., the abnormality appearing in the hybrid individuals). According to Mendelian laws, the normal offspring of affected hybrids when mated with normals should produce all normal offspring; this result is shown clearly through generations IV-VI, where no affected individuals are produced by two normal parents, although one or two of the grandparents were affected. Marriage of a normal person with one affected parent is fit because this individual is wholly without germinal determiners for this character. Marriage between a normal and an affected person is unfit (or it would be if the observed character were a serious defect) because approximately one half their offspring will be affected like the one parent. Thus in IV, 7-21, we see 12 children from one such marriage, 7 of whom are affected, 5 unaffected. All of the 11 children of the 5 unaffected are normal, while of the 16 children of the affected persons, all of whom that married at all married normal individuals, 9 were affected, 7 unaffected. Similar relations are found in generation VI, where the 9 affected persons in V married normals, producing 33 children, 15 of whom were affected, 18 unaffected. Taking all the offspring of marriages between unaffected and affected (hybrid) persons through the four generations III-VI, we find 35 affected and 33 unaffected, with the condition of 3 unknown. There is no instance in this pedigree of the marriage of two affected persons, but such a marriage would be highly unfit (again in the case of a serious defect) because we know that all their offspring would be affected. Mating of two unaffected persons, even though each had one affected parent, would be fit because the offspring would all be unaffected, barring the possibility of a new variation or mutation to this character, which would be extremely unlikely. Such a pedigree as this illustrates very well how a knowledge of Mendelian heredity may be of the greatest value practically, in determining the fitness or unfitness of marriages in families where an abnormality or defect is known to occur. The course of the inheritance here illustrates the simplest form of Mendelism. We have already indicated that there are many other forms which we have not described and which we cannot undertake to describe here on account of their complexity; in such cases, however, it is still possible to predict with fair accuracy the characters of the offspring of parents whose history is known for one or two generations.

The defect we have just been considering is dominant. Many defects are recessive, i. e., transmitted though not exhibited by a hybrid individual. Viewed from the standpoint of the character of the offspring, mating with such a person would be unfit only when both persons were similarly recessives. Such a chance similarity would be likely only in cases of blood relationship. Here lies the scientific basis for many of the legal restrictions against cousin marriage or the marriage of closer relatives, for here, although both persons may appear normal, the chances for latent ills appearing in the progeny in a pure and permanently fixed condition are greatly increased. Of course the same relation holds for characteristics which are not defects but really valuable traits. Marriage of cousins possessing valuable characters, whether apparent or not, might be allowed or encouraged as a means of rendering permanent a rare and valuable family trait which might otherwise be much less likely to become an established characteristic. Some discrimination should be exercised in the control, legal or otherwise, of such marriages.

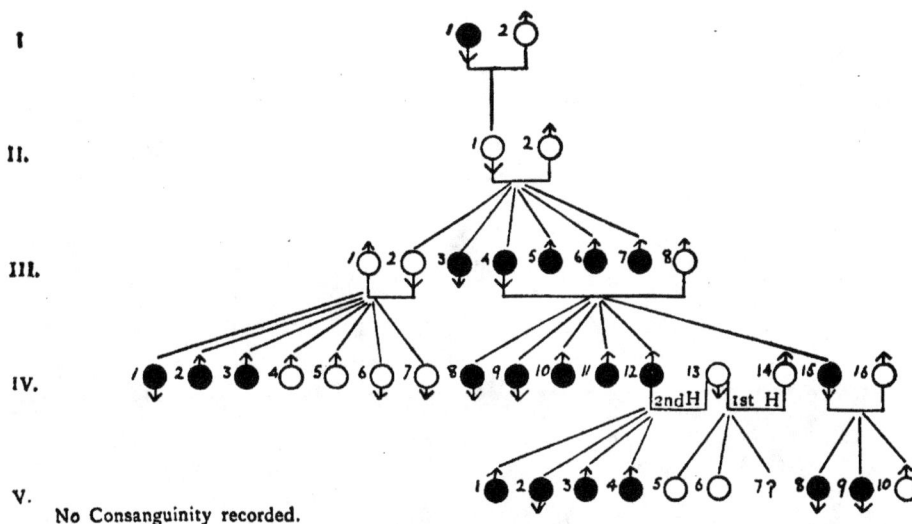

No Consanguinity recorded.

Smith and Norwell's Case.

FIG. 14.—Family history showing polydactylism. (From "Treasury of Human Inheritance.")

Fig. 14 gives a brief pedigree of a family in which polydactylism occurs. This is a condition in which one or more additional or supernumerary fingers or toes are present in the extremities. The Mendelian character of the heredity of this defect is less clear than in the preceding, yet there are many indications that this is really an illustration of a complex Mendelian formula. Probably if the parentage of the individuals marrying into this family were known we should be able to give a complete formula. At any rate the pedigree illustrates the unfit character of the matings with affected persons, for in no instance has such a marriage resulted in the production of fewer than one half affected offspring.

Fig. 15 illustrates a form of what is known as "split hand" or "lobster claw," where certain digits may be absent in the hands and feet. In this case all the digits are absent except the fifth. This is frequently associated with syndactylism or the fusion of the remaining digits into one or two groups. When present this usually affects all four extremities. Two pedigrees of this defect are illustrated in Fig. 16. Here again we have a defect whose inheritance follows quite closely the Mendelian formula, although the character of the matings is not fully known; it is unnecessary to describe the details—the histories speak for themselves.

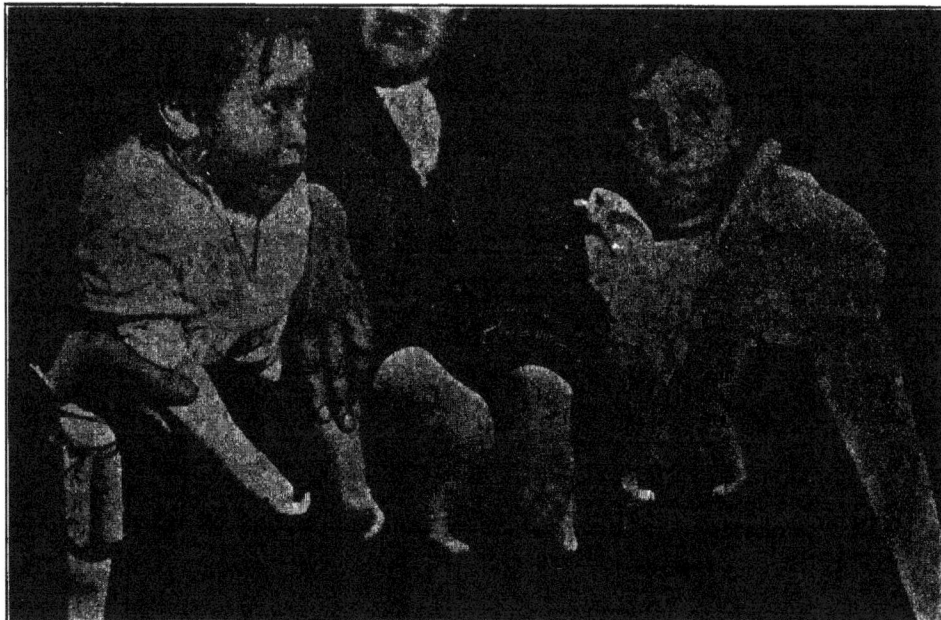

Fɪɢ. 15.—Mother and two daughters showing "split hand." (From Pearson.)

Fig. 17 illustrates a pedigree of congenital cataract. This history is less satisfactory because the matings are given in only three instances. It is known from other data that this defect follows simple Mendelian laws. Normal individuals produce only normals, while affected persons produce one half or all affected offspring according to the character of the mating.

Fig. 18 illustrates the heredity of another defect of the eye called night blindness. This is a retinal defect, the affected being able to see only in strong illumination. The particular form of the disease in this family resulted in total blindness later in life. Little is known definitely concerning the character of the matings; no mating is known to have been with an affected person and some are known to have been with unaffected. Of the 42 descendants of the first affected person only 6 are known to have been unaffected. Can there be any doubt regarding the unfitness of these matings? In generation III a single mating led to a family of 10 children *all* affected by this serious defect, rendering them dependents.

One of the most complete pedigrees of a defect on record is given in condensed form in Fig. 19. This summarizes the extraordinarily complete data of Nettleship covering nine, and in one branch ten, consecutive generations. The defect is another form of night blindness as it existed in a French family. The inheritance is obviously Mendelian: no affected persons are produced by unaffected parents, although their own brothers or sisters or one parent may have been affected. The pedigree gives the history of 2,040 persons, all descended from one affected individual. Of these 135 were known to have been affected, and all were children of affected parentage. Of the total number of progeny of affected persons mated with normals, 130 were reported as affected and 242 as unaffected.

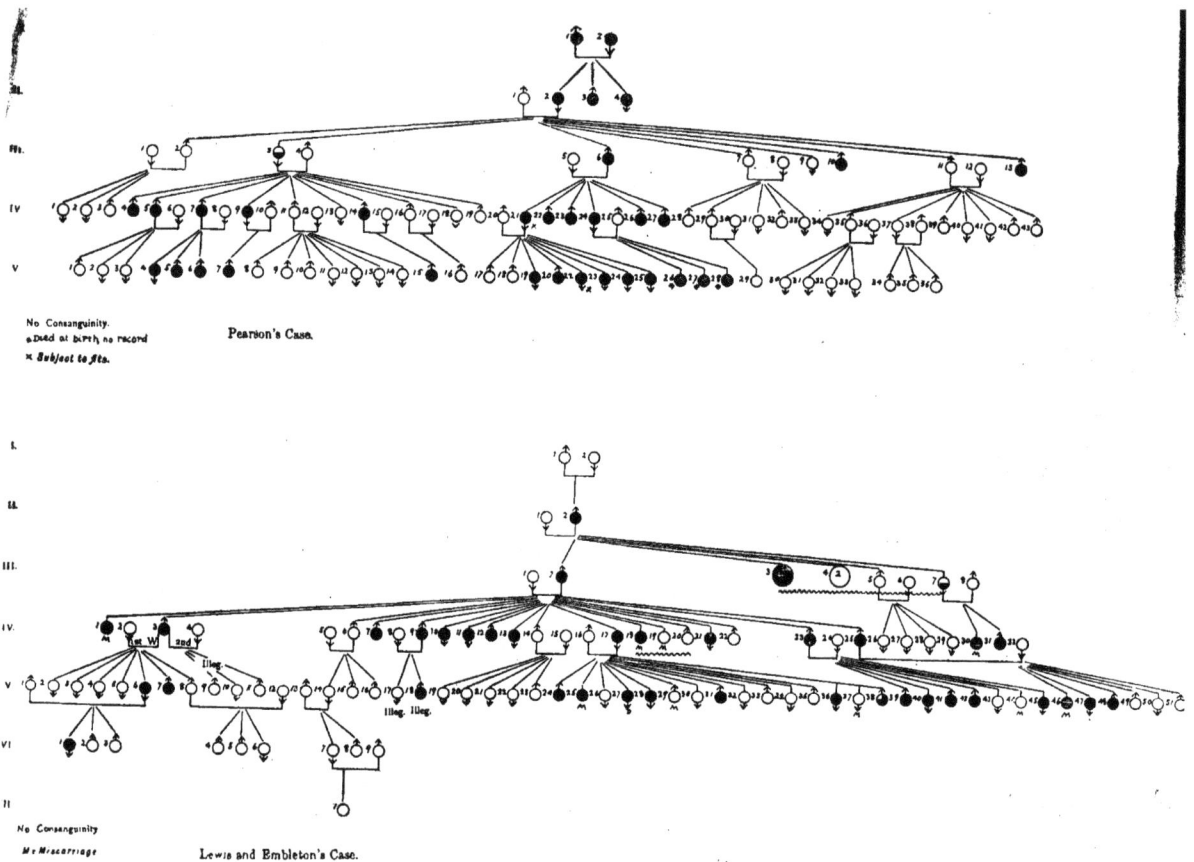

FIG. 16.—Two family histories showing split foot. (From "Treasury of Human Inheritance.")

We may consider next the hereditary history of some forms of nervous defect, the exact nature of the causes of which can be less definitely stated than in all of the preceding instances of defect. Fig. 20 gives a brief history of the heredity of Huntington's chorea—a form of insanity which here resulted in the death of all but one of the affected persons in the first four generations; the fifth generation is the present and is incomplete. Although the matings were with normals in every case, yet in four of the eight marriages all of the offspring were affected. From one affected male 23 affected persons descended in four generations and their multiplication is still going on. There can be no doubt as to the unfitness of marriage into such a family.

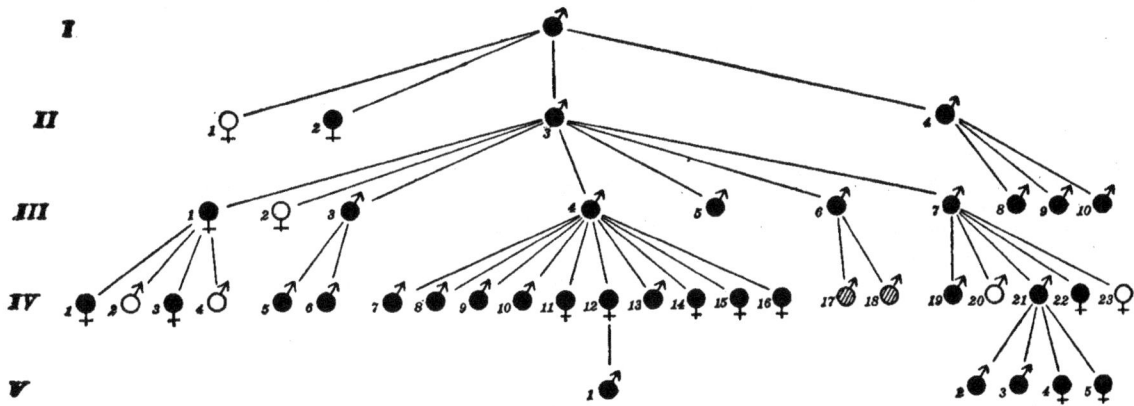

FIG. 18.—Family history showing a form of night blindness. Character of matings incompletely known. (Data from Bordley.)

A very complete family history showing deaf-mutism is given in Fig. 21. It cannot be said that in every case here the defect is innate, i. e., hereditary, and it is not known that the cause of the defect was the same in every family concerned, for deaf-mutism may result from several different causes. In most cases in this history, however, the defect behaves like a Mendelian dominant. In certain other cases it is clearly known to follow the Mendelian formula. Such pedigrees as this show how dangerous it is to marry into a family in which this defect exists.

Goddard has recently published several family histories showing feeble-mindedness. One of the most significant of these—significant both socially and eugenically—is summarized here in Fig. 22. Of this Goddard writes: "Here we have a feeble-minded woman [IV, 3] who has had three husbands (including one 'who was not her husband'), and the result has been nothing but feeble-minded children. The story may be told as follows:

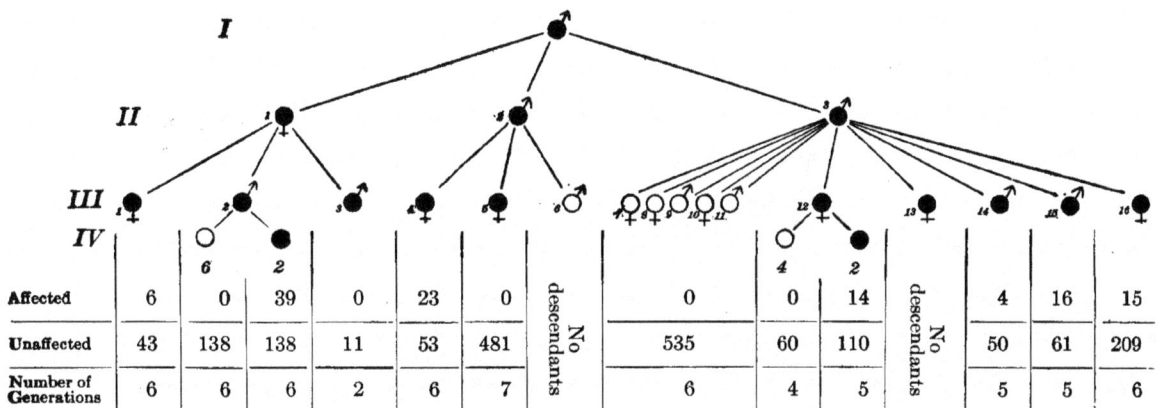

Affected	6	0	39	0	23	0	No descendants	0	0	14	No descendants	4	16	15
Unaffected	43	138	138	11	53	481		535	60	110		50	61	209
Number of Generations	6	6	6	2	6	7		6	4	5		5	5	6

FIG. 19.—Family history showing a form of night blindness. (Condensed form of Nettleship's data.)

"This woman was a handsome girl, apparently having inherited some refinement from her mother, although her father was a feeble-minded, alcoholic brute. Somewhere about the age of seventeen or eighteen she went out to do housework in a family in one of the towns of this State [New Jersey]. She soon became the mother of an illegitimate child. It was born in an almshouse to which she fled after she had been discharged from the home where she had been at work. After this, charitably disposed people tried to do what they could for her, giving her a home for herself and her child in return for the work which she could do. However, she soon appeared in the same condition. An effort was then made to discover the father of this second child, and when he was found to be a drunken, feeble-minded epileptic living in the neighborhood, in order to save the legitimacy of the child, her friends [sic] saw to it that a marriage ceremony took place. Later another feeble-minded child was born to them. Then the whole family secured a home with an unmarried farmer in the neighborhood. They lived there together until another child was forthcoming which the husband refused to own. When, finally, the farmer acknowledged this child to be his, the same good friends [sic] interfered, went into the courts and procured a divorce from the husband, and had the woman married to the father of the expected fourth child. This proved to be feeble-minded, and they have had four other feeble-minded children, making eight in all, born of this woman. There have also been one child stillborn and one miscarriage.

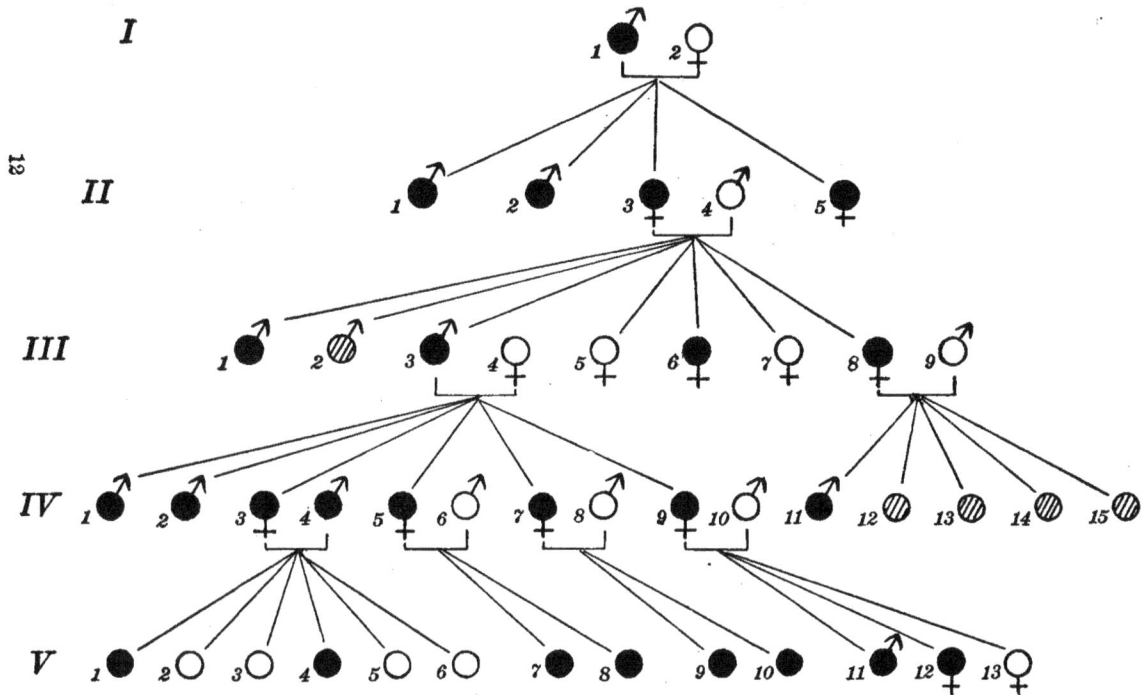

FIG. 20.—Family history showing Huntington's chorea. Last generation incomplete. (Data from Hamilton.)

"As will be seen from the chart, this woman had four feeble-minded brothers and sisters [IV, 6, 10, 15, 16]. These are all married and have children. The older of the two sisters had a child by her own father, when she was thirteen years old. The child died at about six years of age. This woman has since married. The two brothers have each at least one child of whose mental condition nothing is known. The other sister married a feeble-minded man and had three children. Two of these are feeble-minded and the other died in infancy. There were six other brothers and sisters that died in infancy."

The paternal ancestry of this unfortunate woman is hardly less interesting, as may be seen from the diagram. All told, this family history, as far as it is known, includes 59 persons; the mental character of 12 of these is unknown; 10 died in infancy or before their characteristics were known; of the remaining 37, 30 were feeble-minded.

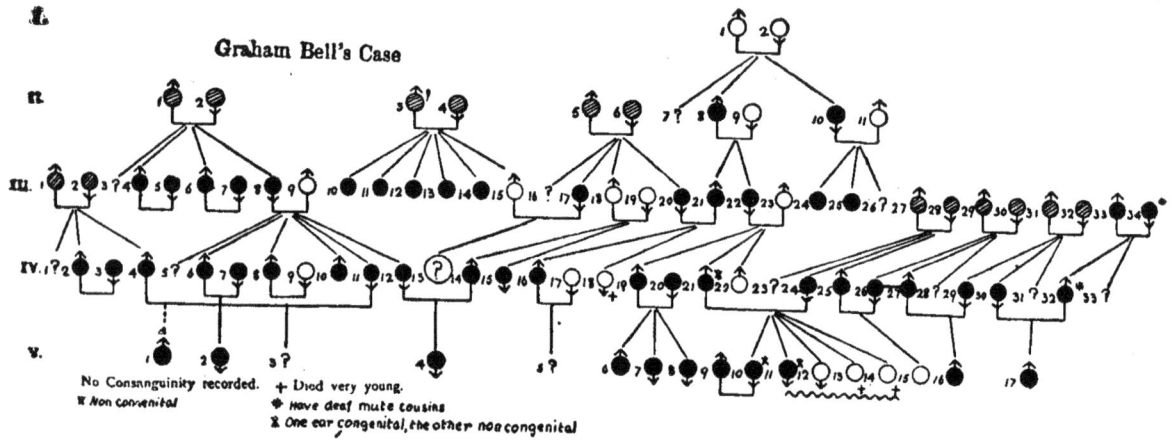

FIG. 21.—Family history showing deaf-mutism. (From "Treasury of Human Inheritance.")

Turning now to defects of other kinds, an interesting history is illustrated in Fig. 23. Here a single individual fatally affected with angio-neurotic œdema gave rise, in four completed generations, to 113 persons, 43 of whom were affected. In 11 this disease was the direct cause of death. The Mendelian character of the heredity here can be neither asserted nor denied. In generations II-V matings between normal and affected gave 42 affected and 35 unaffected offspring.

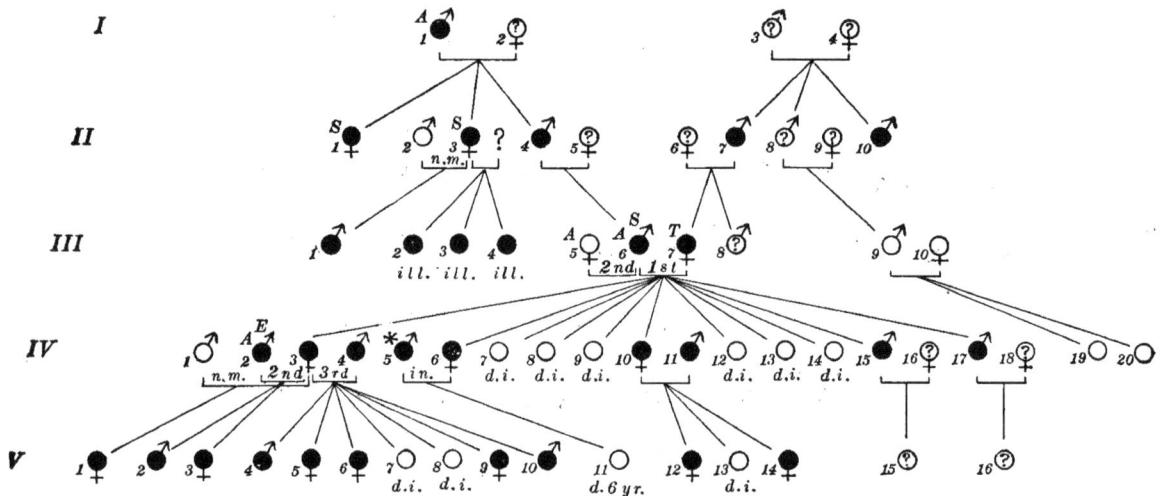

FIG. 22. Family history showing feeble-mindedness. Data from Goddard. *A*, alcoholic; *d.i.*, died in infancy; *E*, epileptic; *ill.*, illegitimate; *in.*, incest; *, same individual as *III*, 6; *n.m.*, not married; *S*, sexual pervert; *T*, tuberculous.

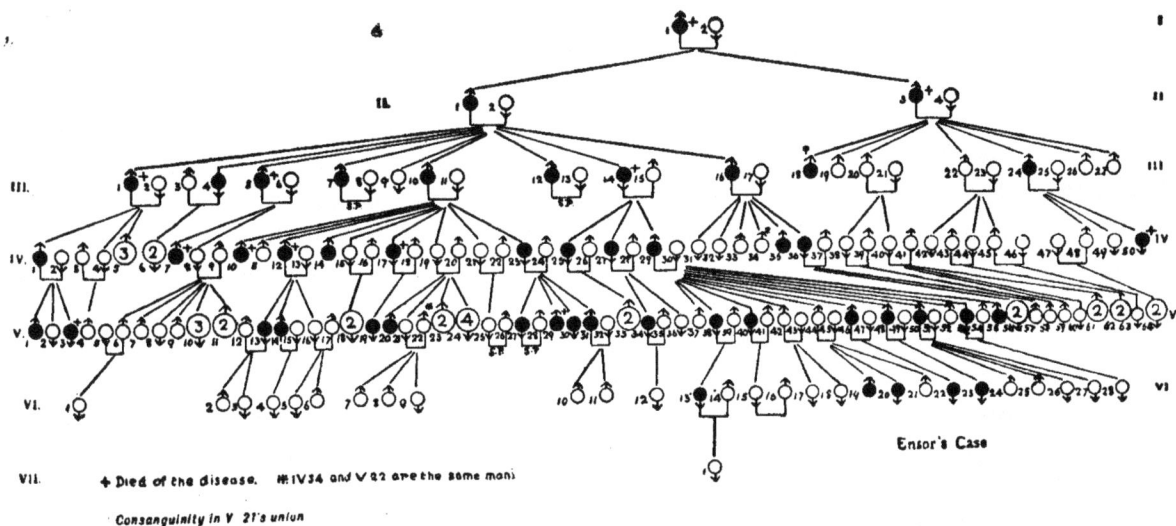

FIG. 23.—Family history showing angio-neurotic œdema. (From "Treasury of Human Inheritance.")

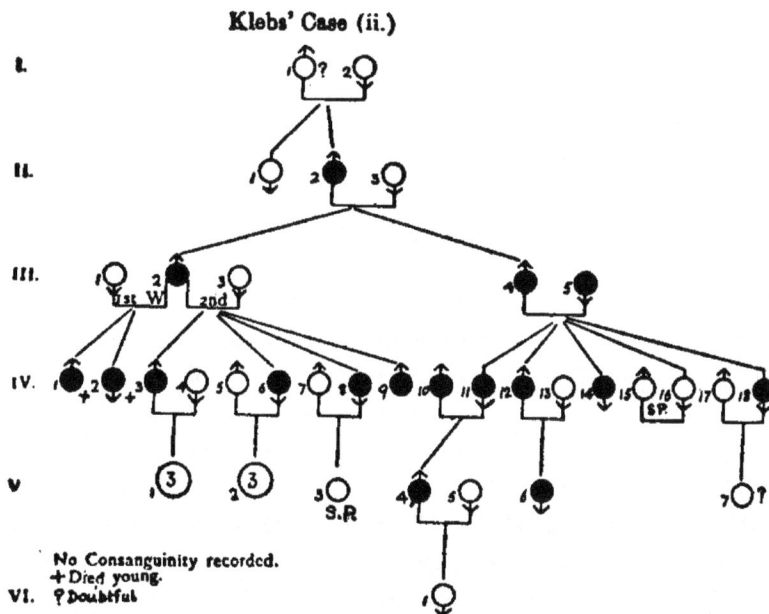

FIG. 24.—Family history showing tuberculosis. (Data from Klebs, after Whetham in "Treasury of Human Inheritance.")

Fig. 24 gives a brief family history showing pulmonary tuberculosis. In the history given susceptibility to this disease behaves as a Mendelian dominant. We cannot as yet say whether this is or is not a general rule. In describing the heredity of diseases primarily due to infection, one or two important cautions must be observed. Of course the source of the infection cannot be "hereditary," and apparently it is only in comparatively few instances that infection occurs during fetal life. To some infections certain persons are susceptible, others are not; some when susceptible are capable of developing immunity, others are not. When an infection is of such character and prevalence that practically all persons in approximately similar environments of a given character are infected, susceptibility or the power of developing immunity will determine whether or not an individual will exhibit the disease caused by the infective agent. Practically all persons living in the denser communities are infected with tuberculosis; those who are susceptible and incapable of developing immunity succumb, the insusceptible and those developing immunity do not. These conditions are heritable; but in speaking of the heredity of such a disease as tuberculosis it should be clear that the heredity concerned is really that of susceptibility and the power of developing immunity. Yet the person who is really susceptible can, by taking sufficient precaution, escape serious infection, and thus the result for that person would be the same as if he were insusceptible, but his offspring would have to take similar precautions if they were to escape the disease.

We cannot speak of heredity in connection with diseases to which all are susceptible and incapable of developing immunity. The presence or absence of such a disease is determined solely by the presence or absence of infection. Many physical and mental defects result from infection as the primary cause. If the infection is one to which all exposed are susceptible and incapable of developing immunity we cannot speak of the defect as in any way hereditary; if the infection is one to which some are susceptible, others not, to which some can develop immunity, others cannot, then we may speak of the defect as hereditary. Thus certain forms of blindness or insanity are due primarily to gonorrheal or syphilitic infection, insusceptibility to which is rare or unknown. Such defects cannot be considered as affording evidence of heredity though they reappear in successive generations.

In general the subject of the heredity of immunity and susceptibility forms one of the most important eugenic aspects of this whole subject. In a few cases it is known that immunity or insusceptibility to specific forms of infection is a unit character which follows Mendelian laws in heredity. It can be added to races or subtracted from them and pure bred immune races built up. So far this has not been demonstrated for man. There is some circumstantial evidence that immunity to specific forms of infection has been a great, although hitherto neglected, factor in man's evolution, and even in the history of his civilization and conquest. It is at once obvious that here is a great field for the common labor of the students of heredity and of medicine and of Eugenics.

Fig. 25 illustrates a family history of infertility. This is apparently hereditary, but before that could be asserted definitely to be so here or in any similar case, we should know that the infertility were not the result of an infection to which immunity is rare or unknown. That infertility is really hereditary in this instance is indicated, first, by the fact that the person marked A later, by a second marriage into fertile stock, had a large family, and second, by the fact that the individual B and his child by marriage into fertile stocks produced in the last generation again a large family and so saved this whole family from extinction.

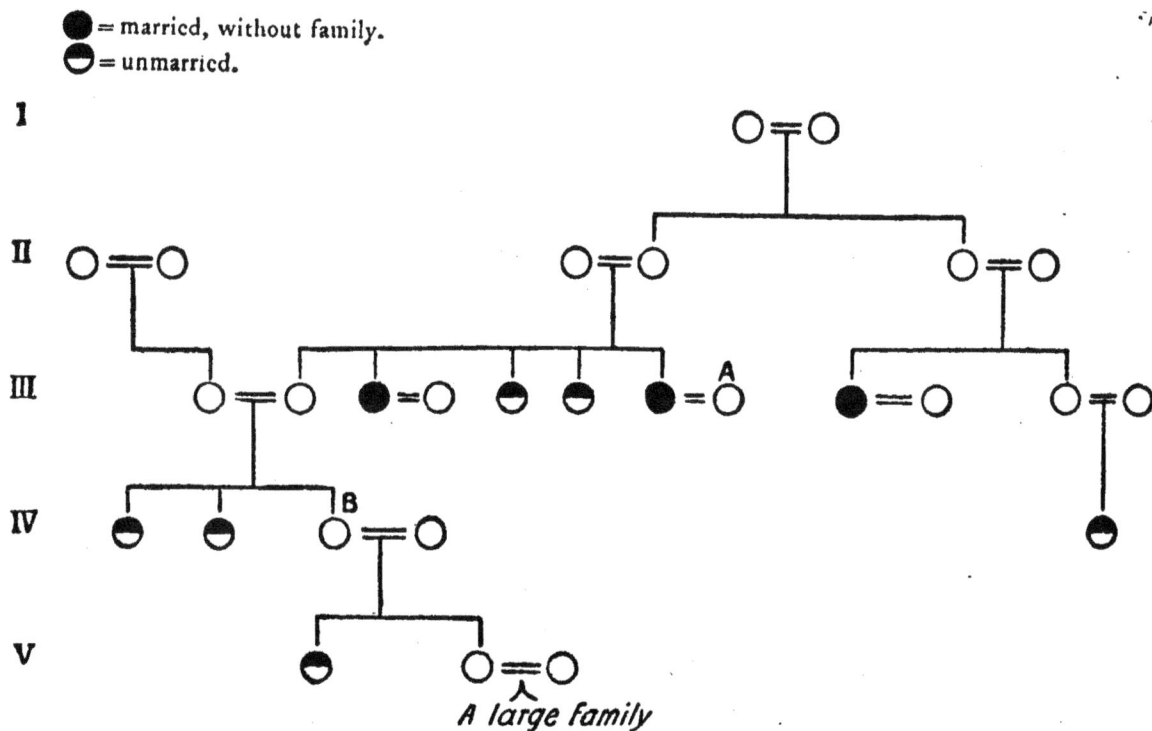

FIG. 25.—Family history showing infertility. (From Whetham.)

Before leaving the subject of the heredity of the kinds of traits we have been using as illustrations, we should add just a word. It is often objected that one cannot properly speak of the heredity of such general things as "insanity" or "deaf-mutism" or "blindness" or "heart disease," because each of these includes a great variety of specific forms of these disorders which cannot strictly, medically, be compared. But the student of heredity replies that when he speaks of the heredity of insanity or heart disease, that is often just what he means. He means that often no particular form of these defects is necessarily strictly heritable as such, but that in a family there may be a general instability of nervous system or circulatory system, which may take any one of several possible specific forms, the form actually appearing depending upon particular conditions which are frequently environmental and beyond determination. In some cases specific forms of disorder are actually heritable as such.

Such an inclusive thing as "ability" may depend upon many different specific conditions. Yet there are families in which persons of exceptional ability are unusually frequent. The fact that persons of ability are more frequent in certain families than in the general population of the same social class and with about the same opportunity for the demonstration of inherent ability, gives evidence of its heredity, although we may not be able to summarize the facts under any particular law but must adhere to their statistical expression.

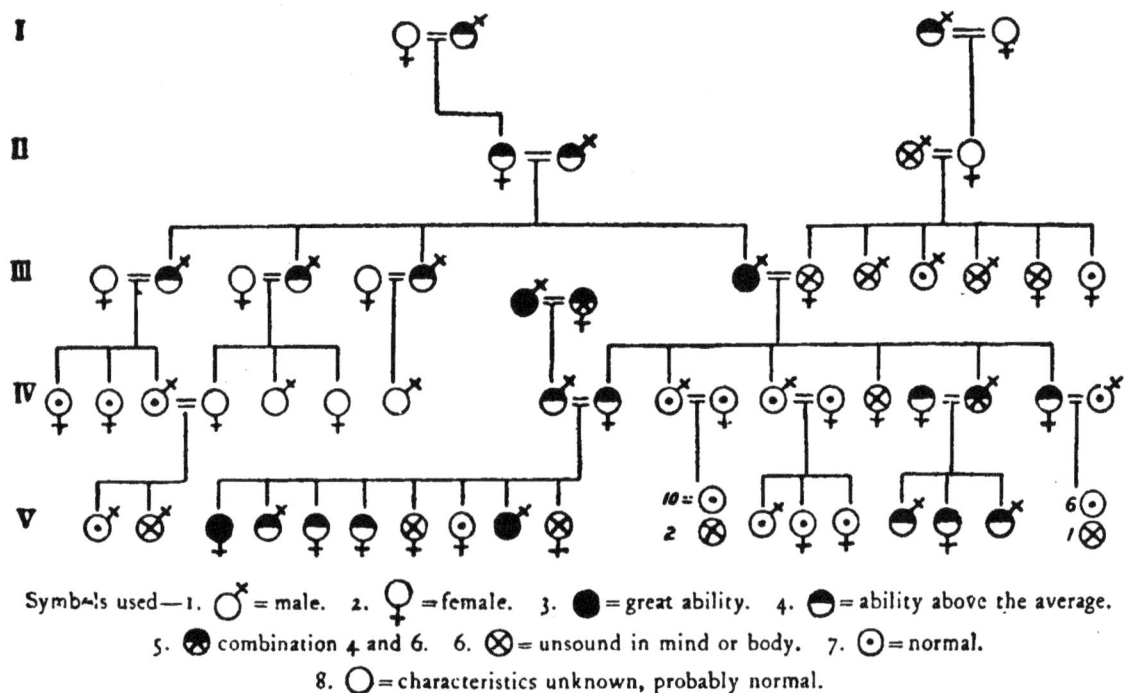

Symbols used—1. ♂ = male. 2. ♀ = female. 3. ● = great ability. 4. ◕ = ability above the average.
5. ✪ combination 4 and 6. 6. ⊗ = unsound in mind or body. 7. ⊙ = normal.
8. ○ = characteristics unknown, probably normal.

FIG. 26.—Family history showing ability. (From Whetham.)

Figs. 26 and 27 illustrate two such pedigrees of ability. In each of these histories there is also a line of "unsoundness" the descent of which it is interesting to trace. It is instructive to compare here the progeny of matings of different kinds. In generation IV of Fig. 26, the 9th and 10th persons are brother and sister. The sister was of considerable ability and married into a family of ability, producing 8 offspring, 5 of whom were able. The brother was a "normal" person and married a similar individual, producing 10 "normal" children. It would be interesting to know the details regarding these two large families of cousins. Another interesting comparison is found in this pedigree. The four able brothers in generation III, coming from a stock of demonstrated ability, married women of undemonstrated ability and all told had 13 children (IV) of whom only 3 showed ability and all of these were in a single family. In this family of the fourth brother two of the able members married into able families, and among their 11 children (second and fifth families in generation V) 8 showed ability; the third able member of this family, however, married as her uncles had, a person not known as able, and none of their 6 children showed unusual ability (sixth family in generation V). Fig. 27 affords other illustrations of this same kind. Thus in generation III the 5th and 7th persons are able cousins of able parentage. The former married a normal and 1 of their 5 children showed ability; the latter married a person of ability and 5 of their 8 children showed ability. In both pedigrees the "careers" of those in the last generation are partly incomplete.

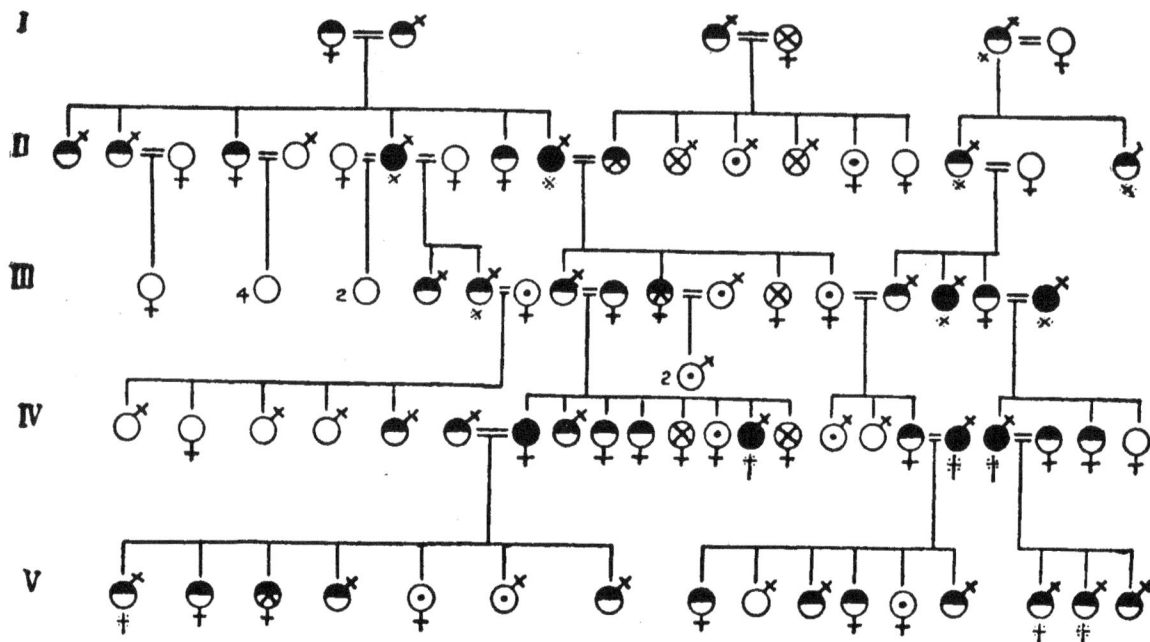

* Biographies in the *Dictionary of National Biography*. † = in *Who's Who.*

FIG. 27.—Family history showing ability. Paternal ancestry of family shown in Fig. 26. (From Whetham.)

In discussing pedigrees of ability it should be borne in mind that the larger proportion of able males as compared with females is hardly significant for the study of heredity; it may merely reflect the unfortunate fact that women have not had the same opportunity to demonstrate inherent ability as have men; or it may evidence the still more unfortunate fact that the distinguished achievements of able women have not been socially recognized as such and recorded as they have been for the other sex.

Fig. 28 gives an interesting, though abbreviated, pedigree of three very able and well-known families. In this history only persons whose ability is in science are marked as able. Charles Darwin is the third individual in the third generation. His cousin, Francis Galton, the founder of Eugenics, is the next to the last person in the same generation.

Many similar cases of the unusual frequency of individuals of musical or religious ability in certain families have been published by Galton and are well known. "As long as ability marries ability, a large proportion of able offspring is a certainty, and ability is a more valuable heirloom in a family than mere material wealth, which, moreover, will follow ability sooner or later."

We might contrast with such families as have been recorded in the three preceding figures some well-known families at the other pole of society. As an interesting example we have the family described by Poellmann. This was established by two daughters of a woman drunkard who in five or six generations produced all told 834 descendants. The histories of 709 of these are known. Of the 709, 107 were of illegitimate birth; 64 were inmates of almshouses; 162 were professional beggars; 164 were prostitutes and 17 procurers; 76 had served sentences in prison aggregating 116 years; 7 were condemned for murder. This family is still a fertile one and the cost to the State, i. e., the taxpayers, already a million and a quarter dollars, is still increasing.

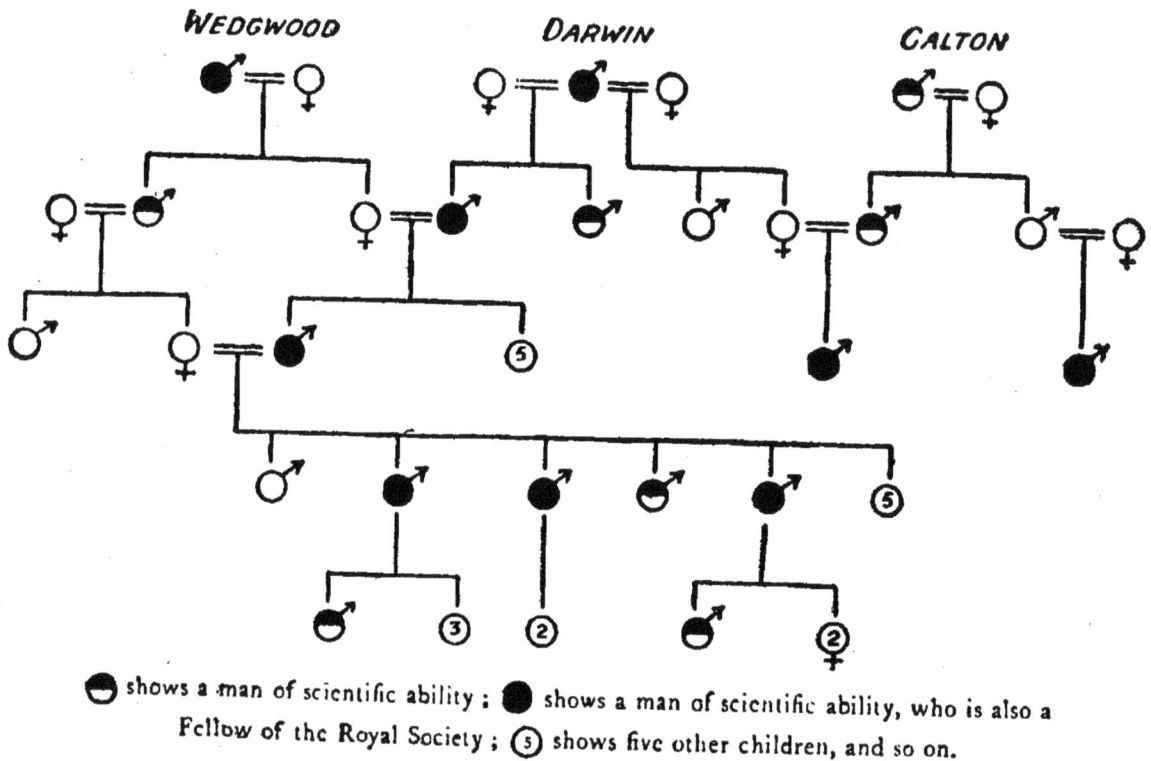

FIG. 28.—History (condensed and incomplete) of three markedly able families. (From Whetham.)

One of the best known families of this type is the so-called "Jukes" family of New York State so carefully investigated by Dugdale. This family is traced from the five daughters of a lazy and irresponsible fisherman born in 1720. In five generations this family numbered about 1,200 persons, including nearly 200 who married into it. The histories of 540 of these are well known and about 500 more are partly known. This family history was easier to follow than are some others because there was very little marriage with the foreign-born—"a distinctively American family." Of these 1,200 idle, ignorant, lewd, vicious, pauper, diseased, imbecile, insane, and criminal specimens of humanity, about 300 died in infancy. Of the remaining 900, 310 were professional paupers in almshouses a total of 2,300 years (at whose expense?); 440 were physically wrecked by their own diseased wickedness; more than half of the women were prostitutes; 130 were convicted criminals; 60 were habitual thieves; 7 were murderers. Not one had even a common school education. Only 20 learned a trade, and 10 of these learned it in State prison! They have cost the State over a million and a quarter dollars, and the cost is still going on. Who pays this bill? What right had an intelligent and humane society to allow these poor unfortunates to be born into the kind of lives they had to lead, not by choice but by the disadvantage of birth? Darwin wrote long ago "... except in the case of man himself, hardly anyone is so ignorant as to allow his worst animals to breed."

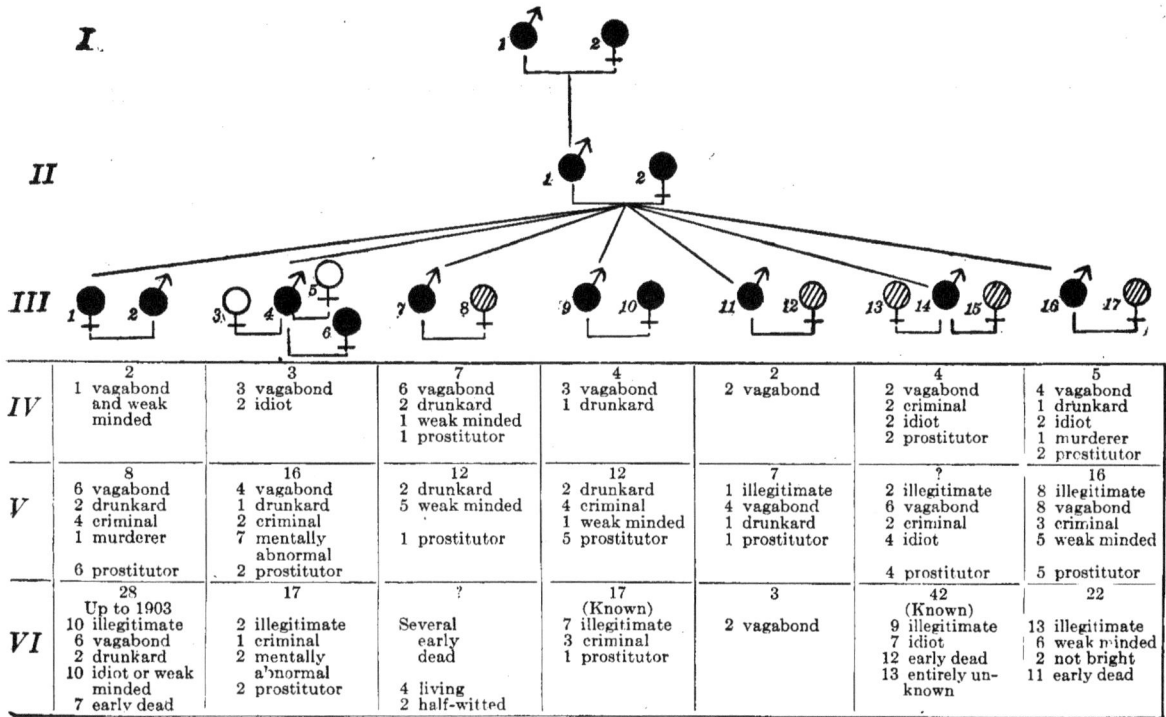

	2	3	7	4	2	4	5
IV	1 vagabond and weak minded	3 vagabond 2 idiot	6 vagabond 2 drunkard 1 weak minded 1 prostitutor	3 vagabond 1 drunkard	2 vagabond	2 vagabond 2 criminal 2 idiot 2 prostitutor	4 vagabond 1 drunkard 2 idiot 1 murderer 2 prostitutor
	8	16	12	12	7		16
V	6 vagabond 2 drunkard 4 criminal 1 murderer 6 prostitutor	4 vagabond 1 drunkard 2 criminal 7 mentally abnormal 2 prostitutor	2 drunkard 5 weak minded 1 prostitutor	2 drunkard 4 criminal 1 weak minded 5 prostitutor	1 illegitimate 4 vagabond 1 drunkard 1 prostitutor	2 illegitimate 6 vagabond 2 criminal 4 idiot 4 prostitutor	8 illegitimate 8 vagabond 3 criminal 5 weak minded 5 prostitutor
	28	17	?	17	3	42	22
VI	Up to 1903 10 illegitimate 6 vagabond 2 drunkard 10 idiot or weak minded 7 early dead	2 illegitimate 1 criminal 2 mentally abnormal 2 prostitutor	Several early dead 4 living 2 half-witted	(Known) 7 illegitimate 3 criminal 1 prostitutor	2 vagabond	(Known) 9 illegitimate 7 idiot 12 early dead 13 entirely unknown	13 illegitimate 6 weak minded 2 not bright 11 early dead

Fig. 29.—History of *Die Familie Zero*. (Condensed from Jörger's data, partly after Davenport.)

Probably the most complete family history of this kind ever worked out is that of the "Familie Zero"—a Swiss family whose pedigree has been recently unraveled in a splendid manner by Jörger. In the seventeenth century this family divided into three lines; two of these have ever since remained valued and highly respected families, while the third has descended to the depths. This third line was established by a man who was himself the result of two generations of intermarriage, the second tainted with insanity. He was of roving disposition, and in the Valla Fontana found an Italian vagrant wife of vicious character. Their son inherited fully his parental traits and himself married a member of a German vagabond family —Marcus, known to this day as a vagabond family. This marriage sealed the fate of their hundreds of descendants. This pair had seven children, all characterized by vagabondage, thievery, drunkenness, mental and physical defect, and immorality. Their history for the three succeeding generations is incompletely summarized in Fig. 29. In 1905, 190 members of this family were known to be living, and probably many living are unknown on account of illegitimate birth.

In 1861 a sympathetic and charitable priest attempted to save from their obvious fate many of these "Zero" children and others who resided in and near his village, by placing them in industrious and respectable families to be reared under more favorable auspices. The attempt failed utterly, for every one of the "Zero" children either ran away or was enticed away by his relatives.

The blame for such an atrocity as this family or the Jukes does not rest with these persons themselves; it must be placed squarely upon the shoulders and consciences of the intelligent members of society who have permitted these predetermined degenerates to be brought into the world, and who are to-day taking no broadly sympathetic view of their treatment by exercising preventive measures. *Laissez faire?*

At the risk of easing the conscience, let us finally return to the other side of society and look at a summarized statement of the Edwards Family given by Boies and drawn from Winship's account of the descendants of Jonathan Edwards. "1,394 of his descendants were identified in 1900, of whom 295 were college graduates; 13 presidents of our greatest colleges; 65 professors in colleges, besides many principals of other important educational institutions; 60 physicians, many of whom were eminent; 100 and more clergymen, missionaries, or theological professors; 75 were officers in the army and navy; 60 prominent authors and writers, by whom 135 books of merit were written and published and 18 important periodicals edited; 33 American States and several foreign countries, and 92 American cities and many foreign cities, have profited by the beneficent influence of their eminent activity; 100 and more were lawyers, of whom one was our most eminent professor of law; 30 were judges; 80 held public office, of whom one was Vice President of the United States; 3 were United States Senators; several were governors, members of Congress, framers of State constitutions, mayors of cities, and ministers to foreign courts; one was president of the Pacific Mail Steamship Company; 15 railroads, many banks, insurance companies, and large industrial enterprises have been indebted to their management. Almost if not every department of social progress and of the public weal has felt the impulse of this healthy and long-lived family. It is not known that any one of them was ever convicted of crime."

The serious consideration of bodies of facts like those contained in some of these pedigrees leads every thoughtful and sympathetic, every humanely minded, human being to ask—What *can* we *do* about it? The display of such conditions stimulates us to measures of relief. It is greatly to be regretted that the honest desire to do good often leads to the performance of ill-considered or unconsidered acts which may result in positive injury to the constitution of society, or at any rate at best merely in the amelioration of the immediate situation without reference to ultimate profit or penalty, or to the necessity for interminable amelioration. Such relief leaves out of account the fact that modifications are not heritable—not permanent, practically without effect in the long run. "Good intentions" have a certain well-known value as paving material, but not as building material.

The science of Eugenics includes not only the study of the data in this field, but further the formulation of definite courses of procedure; but it insists that these be based upon scientific principles and not upon emotional states. Philanthropic relief has become a serious business—is becoming a science. Eugenics is a science and it aims to put the human race upon such a level that the need for philanthropic relief will be less and continually less. We shall then be able to devote more of the resources of our time and money and energy to the production of permanent results. The Eugenist pleads in this work for more sympathetic consideration of the problems of relief—for a sympathy which is wider, which transcends the individual person and reaches the social group, even the nation or race. For just as a society is something more than the sum of its individual parts when taken separately, so the consideration of all the component individuals of a society taken separately and by themselves, results in something less than social consideration. Again "Charity refers to the individual; Statesmanship to the nation; Eugenics cares for both."

What, then, does the Eugenist propose to do? What is the eugenic program? Eugenics is not an academic matter—not an armchair science. It is intensely practical—so very practical, indeed, that the Eugenist hesitates to make many suggestions of a definite nature looking directly and immediately toward specific action. Something must precede action. The Eugenist has been ridiculed as one responsible for the absurd schemes proposed in his name, perhaps seriously, by the unscientific but well-intentioned sympathizer. Many persons have been led to object to what they believed to be a eugenic program which is not a eugenic program at all. Thus the willingness of some to offer adverse criticism of the subject and its aims has grown largely out of a common misconception of the matter and has led Galton to say, "As in most other cases of novel views, the wrongheadedness of objectors to Eugenics has been curious." As a scientist the Eugenist realizes clearly and fully that his new science is in a very early stage of its development. It is just entering upon what are the first stages in the history of any science, namely, the periods of the formulation of elementary ideas and the collection of facts. There are certain groups of facts, however, of glaring significance and undoubted meaning, and upon these as a basis the Eugenist already has a few, a very few, concrete suggestions for eugenic practice. In conclusion, then, we may outline tentatively and briefly a conservative eugenic program somewhat as follows:

First of all there must be an extensive collection of exact data—of the facts regarding all the varied aspects of racial history and evolution. These facts must be collected with great care and under the strictest scientific conditions. In this matter particularly must we "desert verbal discussion for statistical facts." Figures can't lie, but liars can figure. What we need first of all is the accumulation of masses of cold, hard facts, uncolored by any point of view, untinged by any propaganda: facts regarding the net fertility of all classes; facts regarding the racial effects of all sorts of environmental and occupational conditions; facts regarding variability and variation in the race; facts regarding human heredity of normal and pathological conditions, of physical and psychical traits. We have merely scratched the surface of the great masses of such data to be had for the looking. As Davenport has recently put it in his valuable essay on "Eugenics"—

"While the acquisition of new data is desirable, much can be done by studying the extant records of institutions. The amount of such data is enormous. They lie hidden in records of our numerous charity organizations, our 42 institutions for the feeble-minded, our 115 schools and homes for the deaf and blind, our 350 hospitals for the insane, our 1,200 refuge homes, our 1,300 prisons, our 1,500 hospitals and our 2,500 almshouses. Our great insurance companies and our college gymnasiums have tens of thousands of records of the characters of human blood lines. These records should be studied, their hereditary data sifted out and ... placed in their proper relations" that we may learn of "the great strains of human protoplasm that are coursing through the country." Thus shall we learn "not only the method of heredity of human characteristics but we shall identify those lines which supply our families of great men: ... We shall also learn whence come our 300,000 insane and feeble-minded, our 160,000 blind or deaf, the 2,000,000 that are annually cared for by our hospitals and Homes, our 80,000 prisoners and the thousands of criminals that are not in prison, and our 100,000 paupers in almshouses and out.

"This three or four per cent of our population is a fearful drag on our civilization. Shall we as an intelligent people, proud of our control of nature in other respects, do nothing but vote more taxes or be satisfied with the great gifts and bequests that philanthropists have made for the support of the delinquent, defective, and dependent classes? Shall we not rather take the steps that scientific study dictates as necessary to dry up the springs that feed the torrent of defective and degenerate protoplasm?

"Greater tasks than those contemplated in the broadest scheme of the Eugenics committee have been carried out in this country. If only one half of one per cent of the 30 million dollars annually spent on hospitals, 20 millions on insane asylums, 20 millions for almshouses, 13 millions on prisons, and 5 millions on the feeble-minded, deaf and blind were spent on the study of the bad germ plasm that makes necessary the annual expenditure of nearly 100 millions in the care of its produce we might hope to learn just how it is being reproduced and the best way to diminish its further spread. A *new* plague that rendered four per cent of our population, chiefly at the most productive age, not only incompetent, but a burden costing 100 million dollars yearly to support, would instantly attract universal attention, and millions would be forthcoming for its study as they have been for the study of cancer. But we have become so used to crime, disease and degeneracy that we take them as necessary evils. That they were, in the world's ignorance, is granted. That they must remain so, is denied."

Of course one should not jump from this to the conclusion that the fact of heredity is responsible for all of this defect. Disease is so often the result of infections to which none is immune, and defect is frequently the result of such disease. Warbasse has recently stated that "At least one fourth of our public institutions for caring for defectives is made necessary by venereal disease." Doubtless an appreciable share of this fourth is the result of hereditary tendencies, the expression of which gives the opportunity for such infection. Here as elsewhere no single factor accounts for all of the facts, although when, as the result of the increase of knowledge, we shall become able to make more definite statements, we no doubt shall find that heredity is the most important single factor in the disgraceful prevalence of crime, disease, and defect in our communities: indeed this is practically demonstrated to-day. These are questions of the most fundamental importance in our national life-history: our only "hope of perpetuity" lies in the right solution of such problems. And the crying need is for facts, always more facts.

The Galton Laboratory for Eugenics is already doing much in this direction and is publishing in the "Treasury of Human Inheritance" scores of human pedigrees. An agency is already in operation in this country. The American Breeders Association has appointed a Committee and Sub-Committees under highly competent leaders for the collection of exact data of human heredity upon a large scale. There is opportunity for everyone to help in this work in connection with the Eugenics Record Office already referred to.

The second great element in the eugenic program is Research. It is not enough to collect the known facts; new facts must be forthcoming. We cannot, perhaps, undertake definite experiments upon human evolution, but we can and must take advantage of the wealth of experiment which Nature is carrying out around us and before our eyes could we but learn to read her results. We need to know more about the process of differential fertility, of human variability, of the effects of Nurture as well as of the conditions of Nature.

We do know pretty well the effects, upon the individual, of training, education, good and ill housing conditions and conditions of labor, of disease, alcoholism, underfeeding. We need now to know, not to guess at, the effects of these things upon the race, upon human stock. A mere beginning has been made here in the way of a scientific treatment of this question, although many persons have their minds already made up, firmly and fully, as to the "effects of the environment." But all that we have guessed here may be wrong.

The discussion of this subject is filled with pitfalls. The common form of the query as to which is of the greater importance, "heredity or environment," in determining individual characteristics betrays a completely erroneous view of what heredity is, and of the organism's relation to its environment. The living organism reacts to its environment at every stage of its existence, whether as an egg, an embryo, or an adult. In this reaction both factors are essential, the environment as essential as the organism. The result of this continued reaction is the development on the part of the organism of certain physiological processes and structural conditions or characteristics. The nature of these resulting states, depending upon the two factors—organism and environment—can be changed by altering either factor. In general, organisms develop under pretty much the same conditions as their parents and general ancestry did, and their germinal substances are directly continuous, and therefore very similar. Consequently, primary organic structure and environing conditions of development being alike through successive generations, the results of their interaction are alike. This alikeness is heredity—the fact of similarity between parent and offspring. The usually indefinite question as to the effect of the environment ordinarily has a real meaning however, and this is, or should be, whether the alteration of particular elements of the environment, the presence of special, unusual factors which cannot be said to be "normally" present—whether these produce any effect upon the organism which is truly heritable.

This is in reality the old question of the "inheritance of acquired characteristics," or, in a word, of modifications—a question which has been debated heatedly and at length. And as in many similar instances the number of essays and the length and heat of the debate have been inversely as the number and clearness of the pertinent facts. The large majority of biologists have long felt that the great bulk of the evidence was on one side, namely, that acquired traits were not heritable. At the same time they have recognized the difficulty of explaining certain apparently demonstrated contradictory facts. Some recent experimental work has largely cleared away the theoretical difficulties in this field, and the present status of the old and really fundamental question may be stated as follows: External conditions—climate, temperature, moisture, nutritional conditions, results of unusual activity, and the like—incidences of the environment, undoubtedly produce effects upon the structure and behavior of the organism, but these effects must be clearly grouped into two distinct classes.

In the first place the effect of "external" conditions may be to bring about a reaction between the *bodily* parts affected and the environing conditions. Here the body alone is modified and not the germinal substance for the next generation within this body. Such responses to environing conditions do not affect nor involve the structure of the germ, and are therefore unrepresented in that series of reactions that result in the production of an individual of the next generation. In this class are found most of the instances of "functional modification" or acquired characteristics. In this category belong most of the stock illustrations—from the blacksmith's arm and the pianist's fingers, to the giraffe's neck and the fox's cunning. Here also belong the results of training and education; we can train and educate brain cells but not germ cells.

It is characteristic of most of these bodily reactions to external conditions that they are adaptive; that is, when a body reacts to such a condition it does so by undergoing a change which makes the organism better fitted to the new condition—better able to exist. The increased keenness of vision, the strengthened muscle, the thickened fur—all such changes meet new or unusual demands in such a way that the organism has better chances of survival than it would have had unmodified.

But in the second place there are certain environmental circumstances which do affect the structure of the germinal substance within the body of an organism. An unusually high temperature acting at a certain period in the life-history may bring about a change in the color of insects which is heritable—i. e., racial; but such a change results from the action of temperature upon the germ directly and not alone upon the body, which then itself affects the germ. It is essential to recognize that in all such cases it is not the structural change in the body that affects the germ, but it is the external condition itself that affects the germ directly. This is not the half of a hair; it is an extremely important and significant difference. The effects of this kind of action are not visible until the generation following that acted upon. They become expressed in the bodies of the organisms developed from the affected germs.

It is characteristic of such changes as these that they may not, usually do not, have an adaptive relation to the condition bringing about the change. There is no correspondence between the bodily and the germinal modifications resulting from the action of the same condition. Furthermore, there seems to be no adaptive relation between the general character of the germinal disturbance and the environmental disturbance. Rarely some of the organismal characters resulting from such germinal modification may be in the direction of greater adaptedness; usually they are neutral or in the direction of utter unfitness.

But such effects are heritable, whatever their nature with respect to adaptedness, and it becomes therefore very important to find out what are the conditions that may thus disturb the normal structure of the germ. Little more than a beginning has been made here and practically nothing can be said definitely with reference to the human organism in this respect. Enough is known, however, to make it clear that it is only rarely indeed that external conditions can thus affect the germinal structure. In most cases the effects of the incidence of environment are purely bodily. A most fruitful field for eugenic investigation is open here.

One of the first problems to be attacked from this point of view is that of the racial (i. e., heritable) effects of such poisons as alcohol. It is frequently said, for instance, that some of the effects of alcoholism are the weakened, epileptic, or feeble-minded conditions of the offspring, who are also particularly liable to disease and infection. It can hardly be said that this is as yet thoroughly demonstrated. On account of the importance of this question we might call specific attention to some recent investigations of the problem of the racial influence of alcohol. The effects of alcohol upon the individual are fairly well known, although still a matter for debate in some quarters. But this is not as important eugenically as the possible effect upon the offspring of the use and abuse of alcohol by the parents. An investigation has been carried on recently through the Galton Laboratory for National Eugenics directed toward ascertaining the precise relation between alcoholism in parents and the height, weight, general health, and intelligence of their children. It was found to be perfectly true that alcoholism and tuberculosis show a high degree of association; but considering the nondrinking members of the same community just the same high frequency of tuberculosis was found. And the presence of alcoholism among parents was found to be practically without effect upon the height and weight of their offspring. "These results are certainly startling and rather upset one's preconceived ideas, but it is perhaps a consolation that to the obvious and visible miseries of the children arising from drink, lowered intelligence and physique are not added."

The difficulties surrounding investigation and the interpretation of the results of investigation in this particular field are evidenced by the fact that these results have been adversely criticised, on the one hand, because "alcoholism" was taken to mean the continued moderate use of alcohol, and on the other because "alcoholism" was taken to mean only the occasional excessive abuse of alcohol. Much of the confusion surrounding the discussion of the racial effects of alcohol grows out of the underlying confusion of statistical and individual statements. It may be left open, then, whether this result from the Galton Laboratory is clearly demonstrated and whether the basis of investigation was sufficiently broad to make the facts of general applicability.

The frequent association between alcoholism and certain forms of insanity is sometimes taken as evidence of a racial effect. Here again we find the question really left open when we appeal to facts taken in large numbers. In a few cases it seems to have been demonstrated that saturation of the bodily tissues with alcohol affects directly the structure of the germ cells formed at that time, and that this effect is seen in physical and mental disturbances of the offspring derived from such germ cells, and thus becomes hereditary or racial. But these results, like those mentioned above, need confirmation. The impairment of the child *in utero* through maternal overindulgence in alcohol would not necessarily denote any corresponding germinal (i. e., racial) effect.

It is often the case that alcoholic excess, like other forms of excess, may be an indication of a lack of complete mental balance or sanity, sure to have become expressed in some form. The lack of balance in the offspring of such persons is a simple case of heredity and not the result of the parental use of alcohol. The alcoholism of the parent was a result, an indication, and not a cause. There may be instances of the direct action of external conditions upon the germ, and in a very true sense the body is a part of the external environment of the germ, but to say that such an action has been demonstrated for alcohol is premature. It should be easily possible to get real evidence upon this and similar questions. But at present it is safest to leave the whole question of the racial effects of alcohol entirely open pending more and better evidence.

To summarize, then, we may say that the evidence for an inherited effect of the misuse of alcohol is not as clear as one might wish; it may be true. There is the greatest need for the careful scientific investigation of this and allied problems. Much of the evidence here is not of the kind that can be used to prove things—it consists largely of the demonstration of the fact of association rather than of causation. In order to show that a changed environment has produced a change in the innate characters of the organisms affected it must be demonstrated that the organismal change continues to be inherited after the environment has again become what it was originally, and as yet this has not been done. Indeed when tested in this way it is found that a permanently heritable alteration can thus be produced only rarely and by environmental changes of the most profound character.

Research in another direction is greatly needed. We should examine and reëxamine current as well as proposed social practices and reforms from the racial point of view. We should know before going much farther whether the extensive social improvements that are annually effected are to any considerable degree racially permanent. We should investigate not only the racial effects of the unfavorable social conditions themselves, but also the racial effects of the measures directed toward the relief of such conditions. It is conceivable that measures of relief may be practically without permanent effect or even racially detrimental. It would seem that the social worker and philanthropist should welcome any biologically fundamental truths touching these questions, and yet it is curiously true that there are some such persons who seem to prefer not to know the whole truth here, perhaps because they fear it may disclose the unwelcome fact that much of their effort has resulted in amelioration rather than in correction. It should be remembered that simple relief is well worth while, even though often without resulting racial benefit. When it is not actually detrimental racially, relief is an economic, social, and moral duty. The Eugenist, by disclosing the fact that racial effects can actually be accomplished, enlarges rather than diminishes the opportunities for relief and his knowledge should be welcomed and use made of it.

Heretofore the social point of view has been practically the only point of view in much of this work, and the result is that usually following when action is based upon half-truth. David Starr Jordan says: "Charity creates the misery she tries to relieve; she never relieves half the misery she creates," and he goes on to say that *unwise* charity is responsible for half the pauperism of the world; that it is the duty of charity to remove the *causes* of weakness and suffering and equally to see that weakness and suffering are not needlessly perpetuated. In this connection the following quotation from Elderton is apt: "... the influence of the parental environmental factor on the welfare of children is ... at present and has been in the past the chief direction of legislative and philanthropic attack on social evils. Degeneracy of every form has been attributed to poverty, bad housing, unhealthy trades, drinking, industrial occupation of women, and other direct or indirect environmental influences on offspring. If we could by education, by legislation, or by social effort change the environmental conditions, would the race at once rise to a markedly higher standard of physique and mentality? Much, if not the whole battle for social reform, has been based on the assumption that this question was obviously to be answered in the affirmative. No direct investigation has really ever been made of the intensity of the influence of environment on man. To modify the obviously repellent was the immediate instinct of the more gently nurtured and controlling social class. Was this direction of social reform really capable of effecting any substantial change? Nay, by lessening the selective death rate, may it not have contributed to emphasizing the very evils it was intended to lessen? These are the problems which occur to the eugenist and call for investigation and, if possible, settlement.... It is conceivable that the relation between children's physique, for example, and parental occupation is an indirect result of the inheritance of physique and a correlation between parents' physique and their occupation. In other words, what we are attributing to environment may be a secondary influence of heredity itself. A weakling may have no option but to follow an unhealthy trade, a man is a tailor or shoemaker, because he has not the physique for smith or navvy. His offspring may be physically inferior because he is a weakling and not because he follows an unhealthy trade. Clearly, to solve our problem, we must know if there be any correlation between the same character in the parent as we are observing in the child and the environment we are correlating with the child's character. Unfortunately data enabling us to determine the relationship of any mental or physical character of the parent with the environment which is supposed to influence the child is rarely forthcoming."

Just to suggest one further train of thought, we might point out that several movements apparently of high social value have been attended by a curious and largely unforeseen back action. Thus the enforcement of certain forms of Employer's Liability laws has led to discrimination against married persons by large employers of labor and a premium thus put upon nonmarriage. The result of Child Labor legislation has been in some cases an enormous rise in the death rate of young children among the classes concerned, indicating that the children receive less care, now that they have ceased to be a prospective family asset and have become chiefly a burden for many years. In other cases the result has been so serious a limitation in the birth rate that communities are dying out and factories are closing for want of sufficient help. Such problems are not only social but economic and eugenic, and they cannot be seen squarely from any single point of view. It is doubtless shocking to the cultured mind that the chief reason for bringing children into the world should be their economic value as contributors to the family income. But in reality does this point of view differ fundamentally from that very commonly taken of the value of a large family except in the nature of the standard by which their value is measured? May there not be a difference of opinion as to whether children are better or worse off when brought up with some degree of care to be employed under humane conditions of labor, than when left uncared for to die in large proportions of disease and neglect?

Finally, studies in heredity, whether on man or on other animals or on plants, are sure to be of value here because we know that the fundamental processes of heredity are the same in all organisms. Above all, the Eugenist needs to know more of Mendelian heredity in man. The facts of heredity stated in the statistical form of averages and coefficients do not affect the man in the street materially—he rather enjoys taking chances. An extensive eugenic practice can be established only when we can say definitely what the individual or family inheritance will be in a given instance—not what it will be with such and such a degree of probability, although that probability be high. We may not be such a long way off from this ideal, which is an essential for the inauguration of eugenic practice upon a large scale. For the Eugenist this is the richest field for investigation and one which is certain to yield large results.

The Eugenist's demand for more facts will doubtless become an important factor in the progress of biological science. The practical application of the knowledge of heredity in the production of domesticated or cultivated varieties of animals and plants is becoming annually more extensive; and with the recognition of the possibility of the application of this knowledge to the control of the evolution of man himself, will come a rapid increase in biological knowledge and in the earnestness of the student of heredity. And at the same time another result may be that the science of biology shall come to be appraised publicly more nearly at its real value. The biological worker knows that his science comes into contact with human life at every point, that a knowledge of the fundamental principles of the science of life cannot fail to enrich, enlighten, and ennoble the life of every human being. But the community does not yet realize this, to its own great loss. Is it not possible that the Eugenist, finding his fundamentals in biology, by emphasizing the facts of the possibility and the necessity of controlling human evolution, may be able to bring to society a vital sense of the importance of this science with a directness and a vividness which the bacteriologist and hygienist have not been able thus far to realize? Is it even too much to hope that the idea that the "humanities" include only the study of man's comparatively recent past, may now more rapidly give place to a broader conception which shall include not only the whole of man's past, but the study of his future as well? Could any ideal be more vitally, more profoundly human or more worthy of study and devotion, than this of the production of a race of men, clean and sound in mind and body? Be that as it may, the development of this bio-social field can scarcely fail to stimulate strongly the treatment of all social problems with a strictly scientific method. Nothing less than exact methods, and results exactly stated, will satisfy the genuine and really valuable social student of the near future. As one recent writer has feelingly put it: "We have had essays enough."

Eugenic practice for the immediate future is the third part of our program. Must we wait until more data are collected, more facts uncovered, before we undertake any definite proposals for eugenic procedure? Although this is the most difficult aspect of the subject, largely through lack of a sufficiently broad fact-basis, yet we are certainly in possession of enough information to make plain a few necessary steps. Most of the concrete proposals directed toward the reduction of the undesirables and the increase of the desirables have been visionary, impractical, or too limited in their view-point. Above all, they have been open to the objection that they have gone too far in the direction of that zone which separates the two classes. It should be said again that most of these proposals have been those of the amateur enthusiast, not of the seriously scientific Eugenist; they have grown out of that common habit of "getting far from the facts and philosophizing about them."

As Pearson points out, we must start from three fundamental biological ideas. First, "That the relative weight of nature and nurture must not *a priori* be assumed but must be scientifically measured; and thus far our experience is that nature dominates nurture, and that inheritance is more vital than environment." Second, "That there exists no demonstrable inheritance of acquired characters. Environment modifies the bodily characters of the existing generation, but does not [often] modify the germ plasms from which the next generation springs. At most, environment can provide a selection of which germ plasms among the many provided shall be potential and which shall remain latent." Third, "That all human qualities are inherited in a marked and probably equal degree." "If these ideas represent the substantial truth, you will see how the whole function of the eugenist is theoretically simplified. He cannot hope by nurture and by education to create new germinal types. He can only hope by selective environment to obtain the types most conducive to racial welfare and to national progress. If we see this point clearly and grasp it to the full, what a flood of light it sheds on half the schemes for the amelioration of the people.... The widely prevalent notion that bettered environment and improved education mean a *progressive* evolution of humanity is found to be without any satisfactory scientific basis. Improved conditions of life mean better health for the existing population; greater educational facilities mean greater capacity for finding and using existing ability; they do not connote that the next generation will be either physically or mentally better than its parents. Selection of parentage is the sole effective process known to science by which a race can continuously progress. The rise and fall of nations are in truth summed up in the maintenance or cessation of that process of selection. Where the battle is to the capable and thrifty, where the dull and idle have no chance to propagate their kind, there the nation will progress, even if the land be sterile, the environment unfriendly and educational facilities small."

As a concrete example of a most commendable eugenic practice we should mention the sterilization of certain classes of criminal and insane as it is now practiced in the States of Indiana and Connecticut. For the last four years (since March, 1907) the laws of Indiana have permitted the performance of the operation of vasectomy upon "confirmed criminals, idiots, rapists, and imbeciles" after rigid scrutiny of all the mental and physical conditions of the individual case and upon the concurrent judgment of three competent and impartial persons. The title and significant parts of the text of this law are as follows:

> *An Act*, entitled, An Act to prevent procreation of confirmed criminals, idiots, imbeciles, and rapists—providing that superintendents, or boards of managers, of institutions where such persons are confined shall have the authority, and are empowered to appoint a committee of experts, consisting of two physicians, to examine into the mental condition of such inmates.

> *Whereas*, Heredity plays a most important part in the transmission of crime, idiocy, and imbecility;

> *Therefore*, Be it enacted by the General Assembly of the State of Indiana, That on and after the passage of this act it shall be compulsory for each and every institution in the State, entrusted with the care of confirmed criminals, idiots, rapists, and

imbeciles, to appoint upon its staff, in addition to the regular institutional physician, two (2) skilled surgeons of recognized ability, whose duty it shall be, in conjunction with the chief physician of the institution, to examine the mental and physical condition of such inmates as are recommended by the institutional physician and board of managers. If, in the judgment of this committee of experts and the board of managers, procreation is inadvisable, and there is no probability of improvement of the mental and physical condition of the inmate, it shall be lawful for the surgeons to perform such operation for the prevention of procreation as shall be decided safest and most effective. But this operation shall not be performed except in cases that have been pronounced unimprovable: Provided, That in no case shall the consultation fee be more than three (3) dollars to each expert, to be paid out of the funds appropriated for the maintenance of such institution.

This operation of vasectomy, sometimes known as "Rentoul's operation," consists, in the male, in the removal of a small portion of each sperm duct; the individual is thus rendered sterile in a completely effective and permanent way. At the same time there are none of the harmful effects, either physical or mental, such as usually follow the better known forms of sterilization which are in reality asexualization rather than sterilization. Vasectomy is a simple "office" operation occupying only a few minutes and requiring at the most the application of only a local anæsthetic, such as cocaine; and there are no disturbing nor even inconvenient after effects. In the female the corresponding operation of oöphorotomy consists in removing a small portion of each Fallopian tube. In Indiana nearly a thousand persons have already been successfully treated, many upon their own request—a circumstance entirely unforeseen. Similar laws have been passed in Oregon and Connecticut, and are being carefully considered in several other States.

In order that the exact nature of such proposals may be better known generally we may give here also the text of the Connecticut law which is somewhat more inclusive and more flexible than that of Indiana. The Connecticut Statute, enacted in August, 1909, is as follows:

An Act, concerning operations for the Prevention of Procreation. —Be it enacted by the Senate and House of Representatives in General Assembly convened:

Section 1. The directors of the State prison and the superintendents of State hospitals for the insane at Middletown and Norwich are hereby authorized and directed to appoint for each of said institutions, respectively, two skilled surgeons, who, in conjunction with the physician or surgeon in charge at each of said institutions, shall examine such persons as are reported to them by the warden, superintendent, or the physician or surgeon in charge, to be persons by whom procreation would be inadvisable.

Such board shall examine the physical and mental condition of such persons, and their record and family history so far as the same can be ascertained, and if in the judgment of the majority of said board, procreation by any such person would produce children with an inherited tendency to crime, insanity, feeble-mindedness, idiocy, or imbecility, and there is no probability that the condition of any such person so examined will improve to such an extent as to render procreation by such person advisable, or, if the physical and mental condition of any such person will be substantially improved thereby, then the said board shall appoint one of its members to perform the operation of vasectomy or oöphorectomy, as the case may be, upon such person. Such operation shall be performed in a safe and humane manner, and the board making such examination, and the surgeon performing such operation, shall receive from the State such compensation, for services rendered, as the warden of the State prison or the superintendent of either of such hospitals shall deem reasonable.

Section 2. Except as authorized by this Act, every person who shall perform, encourage, assist in, or otherwise promote the performance of either of the operations described in Section 1 of this Act, for the purpose of destroying the power to procreate the human species; or any person who shall knowingly permit either of such operations to be performed upon such person—unless the same be a medical necessity—shall be fined not more than one thousand dollars, or imprisoned in the State prison not more than five years, or both.

These States are to be commended in the highest possible terms for their enlightened action in this direction. Who can say how many families of Jukes and Zeros have already been inhibited by this simple and humane means? "Could such a law be enforced in the whole United States, less than four generations would eliminate nine tenths of the crime, insanity and sickness of the present generation in our land. Asylums, prisons and hospitals would decrease, and the problems of the unemployed, the indigent old, and the hopelessly degenerate would cease to trouble civilization."

And yet probably for years to come those mental states and conditions of servitude graciously termed "conservatism" will continue to insure an undiminished horde of these unfortunates. The situation here is interestingly analogous to that in connection with certain of the infectious diseases. Concerning the eradication of typhoid fever, to mention a single concrete example, competent authorities declare that we now possess all of the information necessary to make typhoid fever as obsolete in civilized communities as is cholera or smallpox. "The average third-year medical student knows enough about typhoid fever to be able to stamp it out if he were endowed with absolute power." "Typhoid fever has passed beyond the catalogue of diseases; it is a crime." Our knowledge of the causes of many of the conditions leading to gross physical and mental defect and criminality has progressed already to such a point that we could if we would eradicate them in large proportion from our civilization. The great horde of defectives, once in the world, have the right to live and to enjoy as best they may whatever freedom is compatible with the lives and freedom of the other members of society. They have not the right to produce and reproduce more of their kind for a too generous and too blindly "charitable" society to contend against. The greater crime consists in allowing the hereditary criminal to be born.

A well-known British alienist, Tredgold, after pointing out that the duty of medical science is to fight and relieve disease in every shape and form, adds: "That if social science does not keep pace with medical science in this matter the end will be national disaster. In other words, I would lay it down as a general principle that as soon as a nation reaches that stage of civilization in which medical knowledge and humanitarian sentiment operate to prolong the existence of the unfit, then it becomes imperative upon that nation to devise such social laws as will insure that these unfit do not propagate their kind.

"For, mark you, it is not as if these degenerates mated solely amongst themselves. Were that so, it is possible that, even in spite of the physician, the accumulated morbidity would become so powerful as to work out its own salvation by bringing about the sterility and extinction of its victims. The danger lies in the fact that these degenerates mate with the *healthy* members of the community and thereby constantly drag fresh blood into the vortex of disease and lower the general vigour of the nation."

Such a practice as vasectomy then represents nicely the eugenic aim of allowing the individual, who is himself never to be blamed for his hereditary constitution, the greatest possible personal freedom and liberty, of allowing full play of sympathy for the individual, and at the same time of exercising the greatest sympathy to society in prohibiting the hereditary criminal from procreating a long line of descendants endowed as badly as he himself was through no fault of his own, but through the gross neglect of society.

Another quotation from Pearson: "To-day we feed our criminals up, and we feed up our insane, we let both out of the prison or asylum 'reformed' or 'cured,' as the case may be, only after a few months to return to State supervision, leaving behind them the germs of a new generation of deteriorants. The average number of crimes due to the convicts in his Majesty's prisons to-day is ten apiece. We cannot reform the criminal, nor cure the insane from the standpoint of heredity; the taint varies not with their mental or moral conduct. These are the products of the somatic cells; the disease lies deeper in their germinal constitution. Education for the criminal, fresh air for the tuberculous, rest and food for the neurotic—these are excellent, they may bring control, sound lungs, and sanity to the individual; but they will not save the offspring from the need of like treatment, nor from the danger of collapse when the time of strain comes. They cannot make a nation sound in mind and body, they merely screen degeneracy behind a throng of arrested degenerates. Our highly developed human sympathy will no longer allow us to watch the State purify itself by the aid of crude natural selection. We see pain and suffering only to relieve it, without inquiry as to the moral character of the sufferer or as to his national or racial value. And this is right—no man is responsible for his own being; and nature and nurture, over which he had no control, have made him the being he is, good or evil. But here science steps in, crying: Let the reprieve be accepted, but next remind the social conscience of its duty to the race ... let there be no heritage if you would build up and preserve a virile and efficient people. Here, I hold, we reach the kernel of the truth which the science of eugenics has at present revealed."

It is also a part of eugenic practice to oppose vigorously and unmistakably any social practice leading to the reduction in the reproductivity of the desirable and valuable elements of society. There is to be included here for censure a long list of customs and practices, from the enforced celibacy of the Church to the horror of horrors—warfare. A moment's reflection will suggest many reprehensible practices of this kind more or less current in certain classes or communities. The requirement of nonmarriage on the part of women teachers—persons of tested and demonstrated ability, is a very general practice of decidedly noneugenic character. In Great Britain more than 75,000 nurses, all of whom must have passed physical examination, are cut off from reproduction by the same requirement of nonmarriage. Many less striking but all too common practices have the final effect of forbidding marriage to the healthy, physically or mentally capable, helpful, classes. "Help wanted. Must be unencumbered."

More vigorously and more unmistakably does the Eugenist discourage anything that leads to matings of the unfit and, above all, to their reproduction. Many countries, from Servia to the Argentine Republic, have statutes forbidding the marriage of the insane, idiots, deaf and dumb, certain classes of criminals, and persons afflicted with certain contagious diseases. It is to be hoped that these laws are enforced with greater effectiveness than that with which our own less stringent laws of similar character are administered. After all, it is the reproduction of these persons that should be limited, and among many of these classes the fact of nonmarriage would provide not the slightest barrier to reproduction.

It is unfortunately true, but true none the less, that there are current forms of so-called philanthropy which, by relieving defective parents of the care of their defective offspring, thus encourage them in the production of more defective offspring; and so the flames are fed. Relief is the smallest part of the problem. Any condition which leads to the multiplication of the innately defective and dependent classes must be sternly opposed. No matter how benign the guise of any form of relief or charity, if it encourages or permits even indirectly the free reproduction of these classes, it must be resolutely opposed and soon abandoned. "It is not enough to preach with horror and indignation against normal parents who restrict their families. Equal reprobation should be the lot of those who, with inherited insanity, feeble-mindedness, or disease, bring children into the world to perpetuate their infirmities. It should not be overlooked that the realization of the power of limiting the birth rate, while it has produced untold harm, when applied blindly and in accordance with individual caprice, may become an instrument for good if it extends to the worst stocks, while the better stocks once more undertake their natural duties."

Practical Eugenics need not be limited to its philanthropic and legislative aspects. There are other social mechanisms which could be used to encourage the multiplication of the fitter, abler families. In Munich, under the enlightened leadership of Dr. Alfred Ploetz, a society for the study and promotion of social and racial hygiene (Internationale Gesellschaft für Rassen-Hygiene) has made a most excellent and significant beginning. This society is doing much not only to collect data and investigate scientifically problems within its field, but also to spread widely the facts of racial integrity. Its members agree, among other things, to undergo thorough medical examination prior to marriage as to their fitness for that state and agree to abstain from marriage, or at least from parenthood, if found to be unfit.

Much can be done by suggestion and suasion regarding the choice of mates and the rearing of large families. When one touches upon this subject he is pretty likely to be met with the objection that the selection of mates is so largely an impulsive, emotional affair that it is quite beyond control. "Marriages," they say, "are made in heaven." But when we consider the number that can scarcely be said to be completed there the statement seems open to some question. As a matter of fact, it is perfectly clear, as Galton,

Ellis, and others have shown, that all peoples, from the Kaffir and the Dyak to the Hindu and the modern European or American, are surrounded with restrictions in marriage often of the greatest stringency. And yet, since these are matters of established social custom, even of religious observance, we submit almost without knowing it.

That results can be really accomplished in this direction and by this method is clearly shown by the history of the Jewish people, and by the Roman Catholics, among whom there are distinctly fewer divorces and childless marriages than among Protestants. In many countries and communities the organized Church still exercises an immense influence over the whole subject of marriage: the Church could easily become a powerful factor in eugenic practice. Such a control can and should be given eugenic direction by the establishment of a more discriminative attitude, looking toward a reduction in the reproductivity of the dependent or defective as well as to the increased reproductivity of the valuable and able. In all of the discussion of "race suicide" and the value to the State of the large family, how seldom do we hear any mention of quality! To plan the organization and conduct of a State without regulating and controlling the quality of its membership is like adopting plans and elevations for a costly building without making any specifications as to materials.

In concrete eugenic practice it seems probable that most can be accomplished for the present by striving to limit the multiplication of the undesirable, dependent, or dangerous elements of the social group. There can be less uncertainty here. The social organization has already marked certain kinds of individuals as unfit and unworthy, whose liberty must be limited in many directions for the social welfare. This aspect of the matter can be put upon a dollars and cents basis very clearly, and this is apparently the only relation that affects a good many people. Why should the able and worthy and thrifty members of society be compelled to pay, as they are in this country alone, $100,000,000 annually, not to mention the vast sums voluntarily contributed toward "charitable" purposes, for the support of the criminal and pauper and defective classes who themselves contribute nothing of value and whose very existence is evidence of criminal disregard of the right of every individual to be well born, into a healthy and sane life? The only answer, if it be an answer, is—because the competent are willing to foot the bill. Millions for tribute but not one cent for defense. And yet a penny's worth of defense outweighs a million's worth of cure.

In the practice of Eugenics the greatest caution must be exercised. All eugenic practice must be tested by the most careful and scrutinizing scientific methods. Mendelian heredity gives a different answer from Job's to his own query: "Who can bring a clean thing out of an unclean?" It also makes clear how it may often happen that it needs but three generations to go from Fifth Avenue to the Bowery, and back again. Many so-called criminals may be anachronisms, some only modificationally bad. But there are many cases, many practices, regarding which there can be no doubt: the Eugenist says, treat these, and let the doubtful cases alone until as a result of the increase of knowledge there is no doubt. And while it is easy to say that we *believe* the criminal or the insane are the products of a wrong environment, it is also easy to say that we believe they are not. What the Eugenist demands is *knowledge*, then belief, and action based thereon.

Finally, the eugenic program calls for the spread of the facts, far and wide, through all classes of society. Bring forcibly before the people the facts of human heredity. Teach them to understand the force of the eugenic ideal of good breeding. "The prevalent opinion that almost anybody is good enough to marry is chiefly due to the fact that in this case, cause and effect, marriage and the feebleness of offspring, are so distant from each other that the near-sighted eye does not distinctly perceive the connection between them." By education we must produce first of all a thoughtfulness in the community regarding the racial responsibilities of marriage and reproduction. Human beings are frequently rational creatures; placing before them clear and truthful ideas regarding fit and unfit matings cannot fail of an ultimate effect. "The virtue of repetition, the summation of suggestion, which sells pills and pickles, which makes Free Trade or Tariff Reform a national issue, this force operating as a slight but persistent influence when linked to eugenic proposals will in a few years' time make these proposals a living force to the common man." By talking and teaching, in season and out, the community will be compelled to think on these things; they will be forced into the public conscience and the pressure of public opinion will rise for the eugenic and against the noneugenic ideals of mating and the rearing of families. And the rest will come in due season and more effective and permanent results will follow than are likely to come from any amount of premature legislation. As Galton writes: "The enlightenment of the individual is a necessary preamble to practical Eugenics, but social opinion by praise or blame constantly influences individual conduct." "Public opinion is commonly far in advance of private morality, because society as a whole keenly appreciates acts that tend to its advantage, and condemns those that do not. It applauds acts of heroism that perhaps not one of the applaud ers would be disposed to emulate." "The first and main point is to secure the general intellectual acceptance of Eugenics as a hopeful and most important study. Then let its principles work into the heart of the nation, who will gradually give practical effect to them in ways that we may not wholly foresee."

In this educational part of the eugenic program, and particularly in the encouragement of research directed toward the solution of eugenic problems and the establishment of eugenic practices, there lies one of the greatest opportunities ever opened to the philanthropist. The genuine philanthropist is he who would at this moment make possible the rapid solution of many of the still baffling problems of human heredity and who would help to spread and teach the gospel of true racial integrity. But while it has been easy to interest philanthropists in the relief of social disorders, few can be interested in the causes at work which make the necessity for relief seem so imperative.

The patient unraveler of the Jukes family history has said, "I am informed that $28,000 was raised in two days to purchase a rare collection of antique jewelry and bronze recently discovered in classic ground forty feet below the *débris*. I do not hear of as many pence being offered to fathom the *débris* of our civilization—however rich the yield!" Possibly one reason for this neglect or omission has heretofore been the lack of evidence that real results could be accomplished in this field. Now that it is so obvious that we have a real foundation of fact from which to work we may expect soon some degree of recognition of the supreme importance of the need for investigation in subjects allied to Eugenics, and of devotion to eugenic aims.

"Whether or no the importance of the issues at stake comes to be recognized fully by the nation at large, individuals and families have it in their power to act on the knowledge they have acquired.... When once more the importance of good birth comes to be recognized in a new sense, ... it will be understood to be more important to marry into a family with a good hereditary record of physical, mental, and moral qualities than it ever has been considered to be allied to one with sixteen quarterings." "Families in which good and noble qualities of mind and body have become hereditary form a natural aristocracy, and, if such families take pride in recording their pedigrees, marry among themselves, and establish a predominant fertility, they can assure success and position to the majority of their descendants in any political future. They can become the guardians and trustees of a sound inborn heritage, which, incorruptible and undefiled, they can preserve in purity and vigour throughout whatever period of ignorance and decay may be in store for the nation at large. Neglect to hand on undimmed the priceless germinal qualities which such families possess, can be regarded only as the betrayal of a sacred trust....

"We look, then, for a day in the near future, when, in some circles at any rate, a comparison of scientific pedigrees will replace, or at all events precede, the discussion of settlements in the preliminaries to a marriage; when birth and good-breeding (in its wide sense), character and ability will be the qualities most prized in the choice of mates; when a bad ancestral strain likely to reappear in succeeding generations will suppress an incipient passion as effectually as it is now cured by a deficiency of education or a superfluity of accent." (Whetham.)

As matters are at present it is all too often the case that marriage is *followed* by the disclosure or discovery of a family history of sterility, or criminality, or insanity. In a truly enlightened society the failure to make known such conditions in the antecedents to a marriage will be regarded as evidence of the greatest moral obliquity, if not of criminal misdemeanor.

The wise and honored founder of Eugenics looks forward to the inclusion of eugenic ideals as a factor in religion. "Eugenics," Galton writes, "strengthens the sense of social duty in so many important particulars that the conclusions derived from its study ought to find a welcome home in every tolerant religion." "Eugenic belief extends the function of philanthropy to future generations; it renders its action more pervading than hitherto, by dealing with families and societies in their entirety; and it enforces the importance of the marriage covenant, by directing serious attention to the probable quality of the future offspring. It strongly forbids all forms of sentimental charity that are harmful to the race, while it eagerly seeks opportunity for acts of personal kindness as some equivalent to the loss of what it forbids. It brings the tie of kinship into prominence, and strongly encourages love and interest in family and race. In brief, eugenics is a virile creed, full of hopefulness, and appealing to many of the noblest feelings of our nature."

And Whetham adds: "Hitherto the development of our race has been unconscious, and we have been allowed no responsibility for its right course. Now, in the fulness of time ... we are treated as children no more, and the conscious fashioning of the human race is given into our hands. Let us put away childish things, stand up with open eyes, and face our responsibilities."

Printed in the USA
CPSIA information can be obtained
at www.ICGtesting.com
LVHW081117160823
755279LV00012B/365

9 781805 479413